Combining

Combining

Nora Bateson

tp
Triarchy Press

Published in this first edition in 2023
by
Triarchy Press
Axminster, England

info@triarchypress.net
www.triarchypress.net

Revised and Reprinted 2023
Copyright © Nora Bateson, 2023
All rights reserved

The right of Nora Bateson to be identified as the author of this work has been asserted by her in accordance with the Copyright, Designs and Patents Act, 1988.

No part of this publication may be reproduced, stored in a retrieval system or transmitted in any form or by any means including photocopying, electronic, mechanical, recording or otherwise, without the prior written permission of the publisher.

A catalogue record for this book is available from the British Library.

Cover Design:
Nora Bateson, Rachel Hentsch, Vivien Leung, Leslie Thulin

Book Layout Design:
Leslie Thulin

Art Work & Design:
Nora Bateson, Rachel Hentsch, Vivien Leung, Leslie Thulin

Wallpaper:
Mats Qwarfordt and Trevor Brubeck

Print ISBN: 978-1-913743-85-7
ePub ISBN: 978-1-913743-86-4

Dedication

This book is offered to your pheromones that help you fall in love.

I give this book to the "you" of so long ago—that felt alone in not fitting into the world people called "normal."

I offer this work to your one crazy eyebrow hair with a twist of its own.

This book is for your breath coming and going unannounced, making you implicit in the same wind that moves the clouds and the trees.

I dedicate this book to the way in which you know—even as you try to describe a dream—that your words cannot express the vivid realm you encountered there, but you grope for language anyway.

These untamed and unmappable phenomena are reminders of uninvited possibilities to be alive together beyond illusions of control.

To these possibilities, I offer everything I am.

Contents

Acknowledgments..................................xiii
Introduction....................................1
Where Prose Stumbles............................3
To Live in Another Way..........................5
Moths and Butterflies...........................7
Do Something...................................9
Eggs Are Time.................................12
Possibility...................................13
Meet Not Match................................21
Hallway of Hallways...........................39
Moving Edges..................................40
Mama Now......................................42
Juicy...44
The Caramels of Autumn........................45
Un-pick-apart-able............................47
Communing.....................................54
Uncut...55
A Pineapple, Tarantulas, Hummingbird, Spiders and Ants .56
Where Is The Edge of Me?......................59
One Thing.....................................65
Without Shields (The Voice of Change Is Changing)......66
Finding a Way.................................69
Tone..75
Traveling on a Paved Road.....................76
Somehow.......................................77
Stretching Edges..............................78
Self Portrait.................................86
Every Hole Is a Story.........................90
I Love You....................................91
Without Going Blank...........................93
Now...94
What I Learned................................96

It's Fantastic	99
Simultaneously Implicating	103
Life Is Art	109
Seasons Change Everything While Breaking Nothing	109
Symmathesy	110
Reunion	111
Crossing Borders: Families in Motion	113
I Fear a Fear of Fear	122
Cracks and Fissures	123
Just Sing	124
Frost	125
Tacit	126
Wild	127
What Is Submerging?	129
Affection for Life	141
Urgent Mud	142
Untamed	144
Aphanipoiesis	145
It's a Gap	177
Listening to the Listeners	183
Noticing	184
Kinky	185
The Meadow-Verse	186
Creature	202
I Am a Crayon	206
Time in Winter Is Underground	207
Unsilent	208
Marrow	209
New Blank Document	212
Yes	215
Divided We Fall Together	216
For You	217
How Do You Pack?	218
Sacred Communication	219
An Ecology of Assholes	220
The Cringe	222
Rejection	223
Two Bad Questions	224
Something New	225
Swerving	226
A Letter to My Imagination	228
Liminal Leadership	233

Words to be Careful With 241
Glossary. 248
Ideas Are Their Stories 256
Theory Is Beautiful 257
The Reasons 258
Salt and Iron. 259
The Zombie Caterpillar. 261
Bacteria 263
Nocturnal 264
Building an Arc 265
Freak Out and Freak In 267
The Rubric. 269
Lurking Monster 271
What Is Sanity?. 272
Common Sense Is Sense-Making in the Commons 273
Frequency 278
(There Is No Script) 279
Minutiae of the Day 280
In the Fire 282
Tearing and Mending. 283
Unbreakable 301
Who-New?. 302
What Am I not Able to Receive? 303
Surreal 304
Decontextualized 306
Family Is Where We Live. 316
Meeting Double Binds in the Polycrisis. 319
Slow Truth. 329
I Want You to Want Me to Want You 331
Silences. 348
Predatory Skills 349
Harvest. 350
To Live It 352
Ecology of Communication 353
Integrity. 361
Something Has to Matter 364
Home 367
Cupped Hands. 369
References and Contexts. 383
Art Credits 386
Contents (In Alphabetical Order) 389
About the Author and Book. 393

Acknowledgments

After I have cooked a meal, and everyone is at the table, and I begin to hear the "mmmm"—the tinkle of forks followed by a silence that comes over those who are eating—I am so pleased. Later, they may say, "Thank you for dinner." And those words reiterate the "mmmm."

What I really would like to say is, "Mmmm."

This book is like a meal, and it came to be through the love, care, hard work, and the long refining of the crafts of many people. I thought I would write a book about warm data, but that was just a lure to get me writing in many other directions, as I avoided writing the book I thought I would write. Writing *about* warm data is not warm data. The about-ness becomes an abstraction that takes away from the possibility of perceiving the living, changing, and combining lifeforms. So, after six years of hard work, this book snuck up and jumped out of the bushes.

One day, it arrived. The book is "Combining."

I called Andrew Carey, my beloved publisher, who had been patiently allowing me to be a forest growing this book slowly in an ecology of ideas, year after year. The next thing I knew, the project had begun. Leslie Thulin, Vivien Leung, and I were suddenly in a rental apartment in Bath, England, with piles of paper everywhere. Vivien brought her art, through which these ideas are given description beyond words. Leslie brought her indefatigable mind, heart, and laptop. My daughter, Sahra Brubeck-Small, joined by phone with astonishing insights, vital redirections, and unflappable integrity. We immediately realized we also needed the artistic brilliance of our dear friend Rachel Hentsch.

Andrew sat down on the floor in the midst of the papers and famously asked, "What would you like to accomplish in the next few days together?" Leslie, Vivien, and I looked at one another in confusion. We had been working together for a few years with the concepts in this book and had become so at home with not setting pre-determined goals or outcomes that the idea was foreign. "We will start to combine and see where it leads." By the end of that three-day time together, the book was forming. Later, Andrew said it was a remarkable accomplishment that ought to have taken months.

True to the ideas in the book, the book itself came about through following the possibilities that presented themselves—and a lot of hard work.

Combining

We entered into the visuals of this book with a vibe, a concept of warmth and luster that opened up the creative talents of Rachel, Vivien, and Leslie. The imagery and the spreads came to life at my house over the course of a ten-day festival of us reading, thinking, and making. My husband, Mats Qvarfordt, is an artisan who makes handmade wallpaper, which he printed with my son Trevor Brubeck. We made their wallpaper into the cover art and the Meadow-verse, combining poetry, ecology, botany, and art. Andrew said we needed to rethink the idea of what a glossary might be, and Vivien made a landscape to play with the map and territory themes in the book. Rachel dove into the transcontextual detective section and the spread on Stretching Edges. Leslie held it all together, bursting with creative ideas and editing text with me. It is Leslie who put this all together in the program and gave it life. Sahra's editing has been crucial; she keeps me on the ground. This is, above all, an intergenerational project.

The text has also been tended by Phillip Guddemi, who knows the Bateson work better than anyone. Other trusted editors of this work have been Howard Silverman, Gail Kara, Lucas Jackson, Lance Strate, Goran Janson, and, of course, Anna Hanschmidt—who also went on an epic finding mission to gather the many pieces of the book that were scattered around the internet, edited, and has consistently shown up with a cup of tea before we knew we needed one. The entire Warm Data community has participated in my learning over the last decade—there is no room here to thank all of you individually, but you have been an incredible research team.

The overtones and the undertones have been nourished by more than a century of Batesons bringing rigor toward their affection for life. William Bateson, my grandfather, and Gregory, my father, are both present on every page. I will never be able to thank them for their courage in standing up to the institutions that have justified exploitation with mandates for objective, decontextualized, optimized systems. They tried in their way, in their contexts, to stop the runaway destruction that we now face. They may not have succeeded, but they did not stop trying. They never sold out. Neither will I.

Thank you so much, everyone.

Introduction

The pieces in this book are created in many tones and textures and come in all sorts of shapes and sizes. I invite you to start wherever the pages open in your hands and, from there, flip around as your curiosity moves you. These compositions will speak to each other through you and your own impressions.

In my own way, I hope to bring buckets of warmth, context, and life into this strange era. This book is made of many things, in many voices, over many years. It is rigorous in its theory, at times expressed through scientific language, at times through personal vignettes and non-verbal expression. It is intimate, and it is global. It is verbal and non-verbal. The expressions are direct and indirect, spanning contexts that range from parenthood to sexual consent to promises and global politics. There is rage and frustration alongside beauty and awe. I am heartbroken and heart-lifted in these pages. Rage requires fear, love, curiosity, and confusion in an ecology of emotions. As a reader, you may find some pieces too spicy or too sweet, some too bitter, too metaphorical, too intellectual, or not intellectual enough. But they are here to feed each other and you, to ricochet and ping, to resonate and echo. I am aware that this is an unusual approach to entering a practice of perception. But I have taken the risk to offer these pieces side-by-side and let them make their own ecology for you, specific to your contexts.

What reads as surrealism to some may be banal to others. Each clump of words and pictures in the book reaches for the others, connecting in the particulars of your reading. They will do so differently as time passes; new resonances will ring. Although a few of these pieces have been previously published in other places, I never meant for any of them to stand alone. Nothing in an ecology stands alone.

It is necessary to have an ensemble of all sorts of communication to meet ecology—intellectual, emotional, storied, non-verbal, and physical. The communications must be diverse enough to meet the diversity of life. Disgust, humor, earnestness, seriousness, sexiness, pensiveness, silent awe, and gushing appreciation allow for the complexity of being alive to meet the complexity of life. It takes every kind of communication imagined and unimagined to meet this time. It takes complexity to meet complexity.

I invite you into this ecology of communication. Pick the flowers, pee in the bushes, throw the stones, watch the clouds, sleep in the shade, and eat the fruit. I welcome you to wriggle and scratch in these pages and find rest and revolution.

WHERE PROSE STUMBLES

Poetry asks us to look beyond words.
Asks us to seek the kind of blur where clarity
is held in combining ideas.
This is important.
Vital even.
The process nuances perception.
And defies algorithms.
Gobbling up rigid definitions like
crispy salted snacks.

Combining

There will be no community
without first communing.

TO LIVE IN ANOTHER WAY

At first, it appears that it is the parts of the system that must be made better or fixed.

Then, it becomes clear that the system is not in the parts – it is in the relationships between them.

So, it seems like it is the relationships that need to be made better or fixed.

But relationships, it turns out, are made of communication.

And then –
The communication becomes the place to address the needed adaptations,

And then –
You realize what is communicated is not what is expressed or even what is not expressed – it is what it is possible to express.

That is where the limits move, in the combining.

How shall I tend to the premises of what is possible to communicate?

Fig. 1. Nora Bateson (concept & text), Rachel Hentsch & Vivien Leung (design), Leslie Thulin (research & layout). (2023). *Moths & Butterflies Spread*. [digital art].

Fig. 1a. imageBROKER. (n.d.). *Eyed Hawk-Moth, Smerinthus ocellata.* [photograph]. Retrieved from stock.adobe.com.

Fig. 1b. dinar. (n.d.) *From The biggest butterfly in the world Attacus atlas close-up.* [photograph]. Retrieved from stock.adobe.com.

Fig. 1c. mramsdell1967. (2019). *Butterfly 2019-119 // Two Blue Morpho Butterfly.* [photograph]. Retrieved from stock.adobe.com.

These organisms are physically expressing a necessary shift in thinking about action. Their wings depict the eyes of birds of prey, and snake heads. However, their predators are rodents, lizards and small birds. Their markings are a communication to the rodents about the danger of owls. Their bodies are illustrating a message of relationship to relationship in context. This is ecology. This is nth-order.

Fig. 1d. Stockgalp. (n.d.) *Butterflies in the subtropical region of MASHPI rainforest in Ecuador.* [photograph]. Retrieved from stock.adobe.com.

COMBINING

How did the informational imagery of the eyes
of the birds of prey get into the wings of the
moths and butterflies?

Whatever it is ...
"It's not just that and nothing more."

Do Something

You don't do a thing in a living system and get the direct result you may have hoped for. You cannot fix the parts to fix the whole, as you can with a machine.

You do a thing, and then something happens so more things happen, mostly in ways that are impossible to track or correlate. The variables excite the other variables into incalculable storms of consequences and consequences of consequences. It is tricky because these consequences of consequences can unfurl and tear apart relational interdependence, as well as generate possibility. It is no longer possible to count how many changes make other changes in how many contexts and directions. Since the number of contexts and directions is unspecified and large, it gets called "nth-order," which is an awkward term. But at least the word invites less underestimation of recursive, looping, entangling, and always moving, conjoining processes that ecologies are.

Let me bring this closer to home. Families are ecologies, weaving each other into many versions of histories and relationships. For some time, my aging mother has been living with us. My son lives next door with his dogs and his girlfriend. Like all other families, our family system is an ongoing story of our histories and possibilities. My mother cared for her aging mother, and now I am caring for her at ninety-four. Anyone who has been a caregiver knows this work is full of humility and peacemaking with difficult moments.

One day, we had an issue with our plumbing. We called the plumber, but they were backed up with work. Meanwhile, my aging mother had to walk upstairs to use the facilities, which was difficult for her, so she decided not to drink much that day and again the next. She became dehydrated and woke up with terrible nausea that evening, and then she didn't eat much the next day. Her legs swelled up, and her breathing became difficult. It was no wonder that her mood dipped, and she snapped at my son for letting his beloved dogs bark.

My son had been irritated with her impatience with his dogs for some time. It had become a representation of her tendency to criticize him. He decided not to come over if the dogs were not welcome, and that meant she did not get to visit with him. She could not be left alone in this state, so my husband had to cancel his day at his workshop. My husband had been so helpful with my mom; I felt terrible that he missed his meeting, and this situation is taking away from his work. I began to apologize . . . his clients, the dogs, my son, me, my mom's legs and heart.

Combining

Chaos had ensued, seemingly because the plumbing was not working. But the plumber could not mend the issue with the dogs, nor could the plumber fix my mom's circulation, or set right the insult to the clients who had an appointment with my husband which he missed. Even when the pipes work, these other issues would have continued, and their consequences would ripple into more contexts.

Systemic change happens in ways already there but has yet to be in specific configurations; the pathways of emergence are determined by the contextual particularities of the moment. Calling the plumber would only bring mending to the pipes (1st-order), but not the scope of possibilities in the circumstances of the dogs, the grandson (my son), my mother's circulation issues, and the fact that I am eternally grateful to my husband for his help with my mom.

The causality is not in the pipes breaking; it is in the conditions in which the plumbing issue occurred, which highlighted all sorts of situations. In hindsight, it may be tempting to assign a linear causal story to the plumbing issue, but that would obscure the necessary peculiarities of the conditions that were already there. Fixing the plumbing ahead of the trouble would not have prevented these conditions from eventually coming to a head. Still, they would have presented in another form, another apparent sequence, and into different tributaries of possibility.

We knew my mother was dealing with circulation issues; we knew about the dogs and that my husband had a hectic workweek. The ongoing stories continue into multiple contexts; it will never stop. The plumber's delay leaked into multiple households that week—and how my husband continued communication with his disgruntled clients probably shifted how he scheduled the next appointment. Then, there is my learning how to wrap my mother's swollen legs and the increasing medical knowledge that comes with caregiving. There is no beginning or end to this story. We were in a "runaway" systemic issue.

It did not take much for the systemic issues to go into "runaway," as many conditions simultaneously produced events that conjoined and enhanced one another in multiple contexts. Runaway is a consequence of interdependency in a system. Interdependency is how life makes life and how systems break down and fall apart. Remember the picture of the moths with eye spots on their wings that look like birds of prey? Those markings reveal the moths' physical evolutionary entanglement in their ecology at nth-order. The animals that eat the moths are rodents and such; the birds of prey (as depicted in their wings) eat the rodents; this is interdependency.

When the relationships that hold a community, a family, a society, or a forest together begin to unravel—runaway happens. The economic vulnerability of so many households combining with high consumption of sugars, combining with stress and mental health issues, combining with floods and droughts combining with intensifying political polarity, combining with media propaganda, combining with artificial intelligence, combining with several hundred years of exploitation, racism, and unmitigated corporate greed . . . all producing issues that are not symptoms (to be quenched one at a time), but are indicators of nth-order; this is runaway.

What people call "unintended consequences" become exponential in their overlapping triggering. Underestimating this process is probably the worst response, and attempts to address it with 1st-order direct correctives make it much worse. What is needed is an understanding of ecological processes and a response that meets the runaway. Combining in new ways across and among contexts is required in these runaway circumstances—not static, calculated, predetermined solutions. As such, very few organizations, governments, or committees are adept in this ability to address runaway and nth-order change.

It takes a deeply personal, in the bone, in the blood, in the spine recognition of the variables in motion. An abstracted version, an impersonal version, or a seemingly "objective" or professional version will pop into modeling and mapping and lose the necessary tangibility of the details—thus losing access to the realms of possibility to meet them.

You may wonder why this book dips into the personal and intimate terrain so readily when the emergencies of our time require global, political action. The practice of perception of nth-order may sound like a mathematical equation or an intellectual flex, but I invite you to meet it intimately—all day, every day. First, one must acknowledge the runaway phenomena instead of attempting to engineer a solution—this acknowledgment is crucial to finding a way through these situations. Touch it, taste it, hold the combining in your hands and heart—never let it go flat.

The revolution will not be a washing machine manual or a newspaper article.
The passion to be human together has always been more . . . unsayable.

COMBINING

```
EGGS ARE TIME

like seeds
future
past.
change
and it is beautiful
persistent
meta for ...
butterfly eggs.
on a leaf
beginning and beginning ...
```

Possibility

Heinz von Foerster once said, "I shall act always so as to increase the number of choices" (1984, p. 6).

And I would like to say, "*I shall act always to increase possibility.*"

What is possible? What is action?

If an emergency medical doctor is treating a person who has overdosed on opioids, they must act directly with the corrective of a Naloxone injection. The action is not in question. However, the way the medic sees the person makes a difference. If the medic sees the person as a human being who has been caught in the multisystemic tragedy of pharmaceuticals, economies, family histories, and media hype, then the way they look at the patient, speak to or about the patient, and the way they touch the patient will contain that communication. If, on the other hand, the medic sees the patient as a social cost or a lost cause, that communication will also be transmitted through touch, voice, and eye contact. The consequences that are opened or closed as communication may not be immediately apparent or measurable but will resonate through the patient's life in unknowable ways. A shift of perception is an action that, in turn, changes the approach. The tone of action in context alters possibility.

Possibility tends to get hijacked by linear causation and goals. The possibility-pesticide of rationality within the existing system will likely mono-crop the awaiting ecology of potentialities. Possibility is sometimes produced by plans changed along the way. The changes change the pathways that change how the change can happen. By definition, these twists must be unpredicted. Preparation for future events is not a matter of knowing or planning what will happen but of learning to be flexible to respond as needed in context.

The uncanny, the surreal, and the impossible are teeming with jarringly preposterous possibilities. They are also out of reach of the alpha rational voice that will forever claim the familiar *is* the only *way*. The familiar is not the only *way*, but it is, in fact, the way to keep the unnamed, unfound, unseen possibilities in exile. When it is difficult to perceive the connections taking place, it is labeled surreal—beyond where reality is perceived. That which does not make sense can be called nonsense, but nonsense according to whose way of thinking? Some would say the global economy is an illusion of absurd nonsense. Others will demand that it is the core reality around which the other experiences of life orbit. Fredric Jameson noted that (wrongly attributed to Žižek) "it seems easier for us today to imagine the thoroughgoing deterioration of the earth and of nature than the breakdown of late capitalism" (Jameson, 1994, p. xii). Perhaps the more ridiculous, incomprehensible, unreasonable possibility is to perpetuate the familiar? What makes deductive sense from this moment is unlikely to continue to make sense. Being locked to a goal obstructs the ability to perceive other responses. It is much easier to look back and see the causality. This tricky habit falsely implies that one could look ahead with the same eye and cast causality into the future. Along the way, however, there are events: rains come late, babies grow, seasons pass, grandparents die, technology is created, new music is released, species go extinct . . . and each of these shifts opens new portals of possibility; some are assumed to be "good" and others "bad." Possibilities are not static; they shift and reshape in response to the inevitable movement in the relationships in play.

Make a wish.

Notice. Was it a wish for the past or the future? I imagine it was for the future. Wishing for the past is a sleepless looping of if-only: if only I had paid attention; if only I had taken more time; if only I had been more curious; if only I had been more kind; if only I knew then what I know now. From the vantage of now, back then is contextualized differently. Now, I can see where the path was heading; back then, I could not. This is the problem with wishes. The wisher is making a wish from how they think and feel at the time of the wish. Their history and experience beforehand shaped their understanding of what was to come, thereby shaping their wish. It is trapped, frozen in time in the epistemology of the wish-maker, and stripped of transcontextual movement. Fables and fairy tales from around the world have brought warnings of this phenomenon. Be careful what you wish for. . . .

The spooky wish stories always include a character who must choose three boons or one wish—and the character, eager to right some situation, produces a wish as a direct corrective. If they are poor, they might wish for wealth; if they are sick, they wish for health; if they are victimized, they wish punishment on their perpetrator. But it always goes terribly wrong. What appeared to be the issue was not an isolated situation. The old storytellers knew that to manipulate the future based on the fixing of a particular problem would always beget regret. The characters in the stories forgot about or could not conceive of the inherent possibility of movement in their situation.

Nothing stays the same. Some characters in the stories wish for the person they fancy to fall in love with them; this is always a disaster. If someone does not fancy you, not only is it likely for the best, but manipulating them into it is a violation. Likewise, the sudden wealth that a wisher longed for erodes their relationships. Perhaps the illness that was wished away also disrupted the possibility of the healing of someone near them as they learned crucial life lessons from caregiving. It could be that the perpetrator they vilified would have become an ally or learned their lesson from another person's story. The surreal details of the way time weaves the stories of our lives are never to be underestimated. The details hold the multiplicity of context and all the possibilities that the wish for a solution cannot. The wish sabotages possibility, obscuring the many transcontextual ways that things can come to be. Ultimately, the wish-maker will need an infinite source of wishes to keep up with the havoc that decontextualized wishing produces.

But the message here is not to do nothing—the message here is not to get caught in the goal that invisibilizes the unimagined possibility. Tomorrow morning is a vast realm of possibility. Ten minutes from now is a vast realm of possibility. Both are blooming through a tangle of threads behind this moment that are shifting each other. The days may go drearily by, but small shiftings are making small shiftings. A change in the tone of voice in a conversation opens the possibility for humor, and a tiny shift in gesture opens the possibility for several future generations of collaboration.

Possibility moves in blobs of more than linear, more than circular, more than imaginable impressions. Tomorrow is pre-soaked in many stories, all congealing over time.

By contrast, the wish becomes a curse and a lesson in humility of "not knowing better" than the complexity of life. To cast a vision into the future is to synthesize what one knows now, from the limitations of this moment, into a future in which countless minutiae will have changed countless other minutiae, and the wish will be ill-fitted to the realm it lands in. The wishes are always wrong. Not sometimes, always. How often have we looked back and thought how fortunate it is that our dream job or crush from eighth grade did not come through? To cast a vision into the future is to project a sterilized singular, lifeless, flat idea into a world alive with weird, wild combining; it will never fit. Occasionally, an artist or a science fiction writer manages to compose an expression with a modicum of rich enough insight to allow for the surrealism of what is coming.

Wish stories generally pivot on the idea that the wisher should have wished for something less selfish, smaller, and more wholesome. A wish made regardless of consequences, based on the wants of the wisher, is not the same as placing oneself in relationship with and communicating within a larger context—this can include prayer. Those things that nourish possibility nourish relationships in more than direct, 1st-order correctives.

Possibility is nourishing nectar. Every drop is needed to meet the multisystemic stuck-ness, the polycrisis, the tautological overlapping of how all the tomorrows are bound up in tech, economic, and cultural traps. Possibility is imprisoned in the locked boxes of causality that lean on each other and feed from each other. The child must be prepared to cloak its soul, take the hit, swallow the pill, listen to the teacher, follow the path to individualized mature adulthood. There was no way to know, even yesterday, how the details of today would change us. Yet that possibility is caged in plans.

Paying attention in this era requires a loosening. From a global perspective, attempting to stop refugees from entering nations where they may hope to live a better life is a reductionist wish. Goods that richer countries have enjoyed and used to enable their lifestyles for the last several decades, all to the detriment of ecology, economy, and political integrity. Violence and division have resulted. Millions upon millions of people will continue to be forced to move due to environmental, economic, and political upheavals. The way in which people who have had to relocate become "othered" as "those people"—in a sentiment of reductionist dehumanizing—utterly omits the possibility that any of us may need to leave our homes in coming decades. This humiliation is inhumane and dangerously provocative. Making enemies, especially of people in need, is never a good idea. The multi-layered aggregate of causal mess is a looming horror story. It will not be easy to meet these challenges, but the possibility lies in the combining and recombining of minutiae—the meeting between actual people, not the types of people, between real human beings with histories and stories that will find a way to help each other. The model for how to deal with immigration obscures the details that would allow for the possibilities to arrive.

This approach addresses the problem directly, from 1st-order, which is also happening in health crises, tech crises, economic crises, food crises, political crises, and cultural divisions. Drumming up the just-right solution for each of the many contexts of conflict and suffering is likely to result in contradictions and unintended consequences in every direction. The possibility is diminished by the fragmentation, the decontextualization, and the metrics of the solutions that are incapable of accounting for the unforeseen openings.

The apparent stuck-ness of any situation is not static; it is not still. Ever-adjusting processes are holding the stuck-ness in place. Movement is there both in possibility and impossibility—increasing possibility invites new combining and space for them to combine.

Possibility is precious. It is also alive. How might it be to hold possibility as something that visits unexpectedly, something that whispers in foreign languages, something that is sacred?

A dear friend recently had a conversation with his granddaughter in which he noticed that he could see a difference himself. His granddaughter's school had scheduled an upcoming "out of the classroom" activity week, a program my friend noticed he would have scoffed at when his daughter (the child's mother) was growing up. Those were the sort of activities that he previously associated with non-learning. But, now, he saw it differently and mentioned to his granddaughter that the week's activities would probably be the most interesting time of the year for her. In his conversation with the next generation, he made a space for her learning to be more than just classroom academics. She was invited into new landscapes of what learning might be. He also showed her that he, as a grandparent, was learning right alongside her.

They entered a relationship where they could learn together, which introduced the sense of what it feels like to be in intergenerational learning. So much "possibility" was loosened into their lives from this conversation—the possibility of how to be in relationship with other people or what changing your perception can feel like. The ways in which the conversation is becoming possibility for the little girl are undefined, and indeed, they will ripple widely into her life. She may become more frustrated with adults uninterested in mutual learning. She may become sensitive to the sort of adults afraid not to be "right"—she may find many contexts of discovery outside of the classroom and feel authorized to give them her interest. One day, she may have a granddaughter with whom she enjoys learning about life.

The story lives in the details of the relationship between grandfather and granddaughter. His observation of "seeing things differently" contrasts the trite declaration that she "needs to have more time outside the classroom for her benefit." Instead of mentioning his learning, what if he said, "You need to go outside to learn; how many times a week do you do that?" The contrast in the approach is the difference between tallying up out-of-class activities and measuring the optimum out-of-class time for her development versus discovering her own learning through her grandfather as he discovers himself through her. The question would completely re-frame the learning into how to *account for* out-of-classroom activities, not at all in the same order as lifelong intergenerational learning. It is just that easy to open—or to limit—possibility. The approach, the tone, the details matter.

The approach is seeking—not to lock down a particular direct corrective, although those are also necessary at times—but instead to include the nth-order—the ecological habit of change that changes change and keeps changing. There is a realm of "possible" waiting.

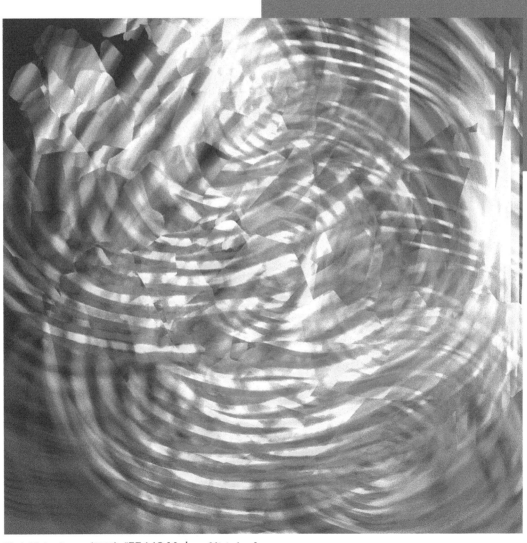

Fig 2. Vivien Leung. (2023). *#EEAAO Madness*. [digital art].

Fig. 3. Vivien Leung. (2023). *Where We Listen*. [digital art]

Combining

Don't be so sure of the way things are.

We are of . . . not about.

Within.

Always.

Meet Not Match

> How does one hate a country, or love one? I lack the trick of it. I know people, I know towns, farms, hills and rivers and rocks, I know how the sun at sunset in autumn falls on the side of a certain plowland in the hills; but what is the sense of giving a boundary to all that, of giving it a name and ceasing to love where the name ceases to apply? What is love of one's country; is it hate of one's uncountry? (Le Guin, 1969, p. 104)

What do you actually want someone to promise you?
Everlasting love? Togetherness? Or?...

The last thing I would ever want is for someone who did not want to be close to me to feel they had to because of a promise. The thought of someone I love deferring their life path on this lovely blue orb so they could continue to fit into a life with me seems to be the opposite of loyalty. Life is shaping us through our histories of different cultures, forming us through the pressures and releases of families, and growing us into and out of jobs, health issues, and projects. Some friends stay close, and others depart. Some rip their affiliation away, others drift. I am the person I was when I was four years old and the one I was at forty, cohabitating with the one I will be at ninety-four. The promise of love is perhaps a promise of attending to those changes, sometimes together, sometimes not. I want the people I love to live their days and find their learning as best they can, with or without me—which is really with me because I want their time being alive to be as it needs to be. In loving them, how could I want otherwise?

The Le Guin question and my questions above invite a loyalty to a larger version of the relationship, and both do so through the forests of particular, shifting specifics—in context. By contrast, the vow or the loyalty to the idea of someplace or someone is an abstraction—this abstracting is perhaps a disloyalty to how the details of life morph and twist into new configurations. Is it possible to be loyal if the loyalty is static? What, then, is one loyal to?

The context of the relationship can require contradicting forms of loyalty. In some relationships, monogamy is loyalty; in others, freedom to be with multiple partners is where the loyalty lies. Is loyalty providing what the other party desires? Or is it loyalty to one's desires? The removal of context, the cutting from dependent relational connection, the objectification that comes with pulling the identified topic, be it a person, a nation, a meadow, or even an idea, from its entangling with other relational entities becomes a violence to the necessary movement that life requires. The abstraction is the decontextualization.

The shape of the response needs to meet, not match, the shape of the trouble

The real question being asked here is not about loyalty but about action. Where and in which direction—can actions be taken that will ripple into consequences of less destruction? The call for and to act is everywhere, ringing out in urgency to do something. There is so much that needs to be done! But action in itself is likely to perpetuate or exacerbate the problems, not solve them.

Action is not always a deed; a lack of response is also a response. A shift in perception may not look like an action but can alter the premises through which action is made. Often, in a room of people eager to address and solve the world's many crises, I am gripped by a panicky sadness. The beauty of the intention to solve these many crises is earnest. Still, the attempted actions are placed in abstracted perceptions that do not meet the rich undercurrents of the situations. The solutions are not where the issue is usually identified—the need to save crops from pests, the need to produce energy, the need to make medicine, the need to make laws, the need to be a respected person, the need to ensure the success of the next generations—and so on. These needs were addressed in ways that undermined ecological, symbiotic processes.

The damage many of these well-intended solutions bring about can be seen everywhere. Most acute problems today are the products of previous solutions. Solutions that were effective responses to a problem at one time (in reductionism) violated unseen interdependencies instead of meeting multiple issues simultaneously. Technical solutions are decontextualized by isolating a problem and creating an efficient, targeted response. So how can an action be taken that attends to the sea of associated contextual possibilities, instead of breaking things?

Alfred Korzybski famously said, "A map is not the territory . . ." (1994, p. 58). This phrase has been reframed into many other versions (but even more are needed).

> The name is not the thing named.[*]
> The menu is not the food.[**]
> The nutrients are not the meal.
> The word is not the thing it represents.[***]
> The representation is not what is being represented.
> The symbol is not the phenomenon it refers to.
> The sign is not what it points to.
> The model is not the modeled.
> The parts are not the system.

[*] (Gregory Bateson, 2000, p. 483)
[**] (Alan Watts, 1995)
[***] (Alfred Korzybski, 1994)

> The message is not the medium.
> The diagnosis is not the person.
> The language of complexity is not the complexity.
> The message is not the meta-message.

The opening to continued movement is in the *is not*—which says nothing of what the map *is* or the territory *is* but leaves an open-endedness. The name is not the thing named, again keeping blurry the possibilities that can continue to unfold. In these statements, neither abstraction is defined in static terms. Nothing is what it is. As Korzybski said, "Whatever you might say the object 'is,' well, it is not" (1994, p. 35).

Making these differentiations may appear to be simple. However, our frequent inability to do so illuminates a muddled place where our habits of perception are trained into the mishaps of identifying systemic consequences as if they were isolated causes. The accompanying habit is to dive into producing solutions to issues that have been misidentified in hopes of generating direct correctives to things that are far too complex to meet in any direct way.

Stop poverty, cure cancer, end war, stop climate change, stop racism, control technology, end inequality, stop corruption—and so on. While I am in absolute agreement that all the above-mentioned crises require urgent attention and must be addressed, the failure to perceive them in their nest of causal tributaries results in responses that only exacerbate the problems and often further distribute their dangers into other contexts. These are what my father would have called *logical typing* errors. In a Korzybskian sense, the identified "problem" is not. And it is unfixable because it "is not."

I have never been entirely comfortable with the term logical typing—it evokes all sorts of programming protocols and hierarchical structures of typology. It turns people off. Logical typing sounds like this was an obscure theoretical artifact of mathematical logic created by Russell and Whitehead, of no use to the rest of us in our daily lives. But nothing could be further from the truth.

The way my father used the idea of logical typing to think of living organisms and ecosystems was very different from some of the connotations of that abstract and austere term. His appreciation for Russell and Whitehead's mathematical set theory became a way to look behind the topic and scan the many relationships that tie the topic into a recursive density of connections. My father, Gregory Bateson, asked, "What pattern connects the crab to the lobster and the orchid to the primrose, and all the four of them to me? And me to you?" (2002, p. 8). This question begins with the study of individual organisms but then shifts into a larger question that includes a patterning of life that is inclusive of you and me (mammals), plants, crustaceans, and life itself.

About and Within

The menu is an abstraction of the food. It is *about* the food—but it is not the food. Eating the menu makes for a flat and papery meal. The meal, by contrast, is interlocked *within* the vegetable garden, the culture of the chef, the conversation at the table, the nature of the occasion, the farmer's children, and the health of the prior and coming generations. The saturation of relations is not the same as that of the menu. Food is much more than nutrients—much more than a menu. While these observations are not new, they are crucial to remember when facing the potential food crises threatened by drought and political upheaval worldwide. If one perceives food in terms of nutrients, then, in crisis, the nutrients will be addressed, not the relationships that make food. In the "about-ness" of an issue, it is possible to address a different dimension of the problem rather than addressing the "within-ness." The conflation of these produces great suffering and an enormous waste of time and money attempting problem-solving where the problem is not.

The problem is what the misidentification leads to. If the identified issue is the menu, that starting point will inform the follow-up inquiry and action. If the problem-solving starts from the *about*, it will proceed in disconnected, abstract ways. To address a food crisis is to address the relational, recursive density that food is contingent upon. It is to nourish the coming together of intergenerational farming, cooking, healing, holidays, ceremonies, culture, seasons, and markets—not to package nutrition bars. Likewise, if any complex system is in crisis, it is necessary to be very careful where the identification of the issue is placed.

What surrounds the nutrients? What is a meal? What is the compost? One prefers to eat something other than the compost. Still, the way of describing the relationship between the compost and the conversation at the table is beyond the reach of familiar causal descriptions. The process of assigning a category involves leaving out details. The more information left out, the more general categories are defined, and the more abstract the description becomes.

Through the study of general semantics in the 1930s, Korzybski began a project that looks at how things happen in life and are perceived, how those perceptions become specified by language, definition, and names, and then how that language is leveraged into a sense of control of how we think, feel, and act. Several hundred years of developing categories—and mechanistic ways to control, name, and measure those categories—have set most of us at odds with the ways life makes more life. The results of this confusion are seen everywhere, from education programs to jails, to war, to food, to sex, to identity. This confusion permeates our most intimate ways of thinking about and perceiving health, family life, love, spirituality, gender, law, and so much more. Korzybski said, "Let us repeat the two crucial negative premises as established firmly by all human experience: (1) Words are not the things we are speaking about; and (2) There is no such thing as an object in absolute isolation" (1994, pp. 60–61).

Because Russell and Whitehead's term "logical type" or "logical level" originates from mathematical set theory, it still smells like the kind of math that scares people. While the history of the idea is vital to understand, the concept itself is too important to leave stranded in an intellectual zone that leaves so many people out—to merely describe the importance of this as a familiarity with set theory gets us into discussions about honing attention to the difference between a set of furniture and a set of chairs. Chairs are furniture, but not all are chairs (Whitehead & Russell, 1910). Furniture has a membership of more relation to how people use and make furniture, including tables, sofas, bookshelves, desks, armoires, and cabinets. These relations lead to and are produced from a recursive density of connections to the things we do, the things we collect, the time we spend alone or with others, the way we relax, the way we work, etc. The reach of this is far beyond chairs and furniture. It is far beyond menus and food.

My father was more likely to discuss logical types than to use the word "levels"—but not exclusively. The notion of hierarchies lurks in comparing the map to the territory, the menu to food, and the name to the thing named. There is an implicit contrast that places one in a "higher" saturation or density of relational, recursive processes than the other. This noticing is critical to the practice of perceiving in this way, yet it can also cause further confusion. The lion may appear to be the highest organism of the savanna, but without the bacteria in the soil under the grasses, there is no savanna—and no lion. The types can flip depending on context. A "leader" may appear to have social and political power but is only given that position by their followers, whose adoration may flip disdain in a heartbeat. This is not a rubric of definition, which makes it confusing to explain how to know which abstraction is which.

The other confusion around assessing differences in abstraction is that the confusion itself is destructive and creative. The moment it becomes apparent that perceiving these differences is necessary for anyone hoping not to destroy the world, it is crucial to address the fact that the opposite is also true. The mismatched gaps in contextual processes sometimes produce humor, art, play, learning, and even evolutionary leaps. The rubbing together of different abstractions can create epistemological insight, unprecedented possibility, and even new life forms. It is all the more reason to hone this perception but to recognize that it is rational and non-rational. Try as one might to lock down a methodology for how to grasp this; it seems in the end that it is something physical: the belly knows. Again, this is not very helpful in setting up a "way of knowing,"—especially while trying to make a case that there is great urgency to gain this sensitivity. I will keep trying. That is the practice.

My father used the example of puppies communicating to show the difference between a bite that is a real bite and a bite that is a play bite. The play bite contains a meta-message, communicated through other gestures and tones to invite a romp of non-fighting, conveying that this bite is not a fight bite but play. Play among puppies, humans, or other organisms is often a mixing of communications—abstractions that loosen potential and practice for learning. Play and humor point to the limitations of perception and poke them. "Gentlemen, you can't fight in here. This is the war room" (Kubrick, 1964).

If you listened to a piece of music and assessed that piece of music by how many "A" notes or "B" notes were played, you would miss the music. You would miss the music if you piled up all the rests and rhythmic info. The music is in and between the notes, rhythms, and silences, but it is also in the memories of your life, the impressions the music brings up for you, the heartbreaks, and the longings of your existence. The music meets you in a different way than the information about it does. While this may seem obvious, to extrapolate this idea into social, economic, and legal systems is to notice that this kind of broken assessment happens all the time. Worse still, the solutions to the issues are misinformed by broken, abstracted information. What is a written score? Where is the resonance of the music?

It takes a while to allow this seeking of relational, recursive density or saturation to be the first glance at a circumstance. But it is not impossible. I wish the schools would focus on more opportunities for eight-year-olds to learn this. This way of perceiving a person, a social issue, a climate crisis, a legal battle, or a technological innovation applies to all of us—every day, all day. To say that learning logical typing is a survival skill is about right.

What shall we call it if not logical typing?

I don't know; perhaps a category error, a labeling error, or an abstraction error. But the name is not the thing named, which is already implicit in the theory. "Pig" is a name for an animal, but when I use the word, there are no snout-snorting animals on the page. I cannot eat the bacon from the word pig. They are "pigs" in very different and essential ways. Nor can I think about pigs and have actual beasties running around inside my head. One is not better than the other, but to notice the difference is a radical thing to do.

I remember my father talking about these pigs in his head when I was a kid. I was amused at the idea of little piglets jumping around in his head. It seemed so silly at the time that he thought this to be such an important insight. It was obvious to me as a child that pigs don't live in people's heads. But soon enough, it became clear that this pig problem was not just about pigs. All I had to do was get in trouble for breaking a rule that could not hold the context of why I broke it to realize how many pigs were running around in people's heads. "But that is a stupid rule," I would exclaim. "Yes," said my father, "most rules are." The sorts of shifts that could come about if this concept were truly understood would be profound. The problems come when the reaction is to the word, as though it were the thing. While this may sound obvious, it is a prevalent problem. How often are people referred to by a decontextualized description: a diagnosis, nationality, skin color, sexuality, and so on?

Russell and Whitehead first spoke of logical levels/types, and my father brought this into his work with communication, ecology, and the study of life. Korzybski did as well. While they are both renowned academic philosophers, they are not alone in understanding this vital mishap of perceptions. Marshall McLuhan teased Robert Browning's line, "A man's reach should exceed his grasp, else what's a heaven for" (Browning & Markham, 1856, p. 187)? He gave us a double dose of the need for reaching into those aspects of communication that are ungraspable and thus make room for the new, for play, for life, "A man's reach must exceed his grasp or what's a metaphor?" (McLuhan, 1964, p. 7).

And what is a meta for?

The mishap flips around on itself. Making abstractions into non-abstractions and vice versa—people find it tricky to tell which is which. It is strange to notice how discussions around living systems in interdependency are often received as impossible abstractions because of our familiarity with a world of divisions. But, of course, nothing could be less abstract than the relational realms of life. It is the categories and compartmentalized fragmentation in our thinking that are abstract! The known landscape has become the premise; by contrast, the infinite combinings of living things evade graspable language and thus seem very abstract indeed. When something is not understood, it is commonly described as abstract. Perhaps the term is an indication that the net of relational context around the topic is obscured—not perceiving those relationships may be what produces the sense of abstraction. Sometimes music, art, or poetry can feel abstract at first, and later, as the connections and scaffoldings they are composed within become more evident to the audience, the abstraction becomes a grounded sense of experience. If one does not know the mythologies of the illustrations painted on the walls of St. Peter's Basilica, the characters and their depictions seem abstract. If one recognizes the stories, the same images evoke and comment upon a rich tapestry of narratives. If you are unaccustomed to the logic of a culture you are visiting, the grocery store layout may seem random. But, with an understanding of other aspects of the culture, the placement of the items in categories, unlike the ones at home, falls into place. Which abstraction is the abstraction? Is there a way to explain the difference? The unfamiliar is experienced as surreal, abstract, disconnected, and disorienting—precisely because it is not oriented into one's own encultured and recognizable architecture of life.

My father used the art of storytelling to ground some of the concepts he was exploring that are inconsistent with the habits of perception in an industrial world. He would often use the same story to illustrate many concepts. These stories were usually peeks into communication and learning in unexpected ways. One story from his time in Hawaii, where he interacted with porpoises, has been told repeatedly. Sometimes, the story addresses his concept of the double bind, and in this version, he uses the same story to show a shift in logical typing. In this story, there are many substories as my dad pokes at the strange (il)logic of the way the human beings have entered into interactions with other species in a world where the researchers need the "fish" of ticket sales and find themselves in the ironic act of creating performances in amusement parks. Who is the clever creature here? Who is the trained performer?

THE PORPOISE STORY—LOGICAL TYPING VERSION

> Let me give you an example from the animal level. When I got to Hawaii, I found a performing porpoise establishment which tried to get its money by selling tickets for porpoise shows to tourists. And they hoped to get enough money that way and be able to do some research; it didn't work out. Anyhow, the trainers had been allowed to plan the performance for the public, and they had a porpoise in sort of a holding tank in the background and a big glass transparent tank in the foreground in which the porpoise is going to perform. And the trainer would say to the audience, "We are going to show you how we train porpoises."

> Now, they believed with some innocence that they were, in fact, using operant conditioning to train porpoises. That's Skinnerian psychology: that the act which is in some way rewarded will be repeated. And they say, "Well, we have trained this porpoise already, that when she—she was a female—when she hears the whistle, she will come over to me, the trainer, I will give her a fish, and then she will go and do again that which she did to get the whistle and the fish." And indeed, this was happening quite nicely, and the public was being shown how they trained porpoises. But there was a sort of difficulty. You see, at the next show two hours later—there were five shows a day, six days a week—at the next show,

it wouldn't be right, would it, to teach the porpoise the same thing that you've taught it in the first show because this would be faking, you see. So, the porpoise had to do something different in the next show. And, in fact, the porpoise had obviously learned that it should do a new thing every time it came on the stage—a very high-grade piece of learning. I spotted this and got the outfit to record the training of a porpoise for this performance.

We started with a "naïve" porpoise from the ocean, trained her to do the whistle and fish bit, and then set up learning situations which would run about ten minutes and the rule was that in no learning situation should the porpoise be rewarded for anything that it had been rewarded for in a previous learning session. It could only be rewarded for doing something new. So, the porpoise comes on stage—this is session one—and starts to cruise around. Nobody seems to be paying any attention. The porpoise does a head flop to say, "The hell with it, where is everybody?" or something, and the trainer regards that head flop as a piece of behavior, blows the whistle, porpoise goes and gets the fish, and the porpoise then does the head flop again, gets another fish, does the head flop again, gets another fish, and that's fine. Right. Next training session, porpoise comes on and does its head flop and expects the whistle and fish; no whistle and fish. Does it again; no whistle and fish. Does it again; no whistle and fish. Wastes two-thirds of the session time doing what had been rewarded in the first one. Then becomes angry, gives a real good tail flap, which again the trainer takes as a piece of behavior, rewards, etc. Third session, the tail flap wastes two-thirds of the session, and some new thing comes in more or less by accident at the end. And this goes on for fourteen sessions, in which the porpoise is wasting two-thirds of every session not getting food by doing the wrong thing—doing the thing which had been rewarded in the previous session.

Between the fourteenth and the fifteenth session, the porpoise was put back as usual into its containing tank—kennel, whatever you want to call it—and there became almost insane with excitement—double flipping, all sorts of screw movements, twisting movements, and so on. And when finally led on stage in the fifteenth session, she did twelve absolutely new pieces of performance, one after the other, but was, of course, only rewarded for one of them because that was the rules of the game that the experimenters had set up. Now, you see, the porpoise was in a tangle with a false belief: the false belief being that what was rewarded in session "n" is what should be rewarded in session "n + 1." The porpoise had a problem of getting a perception—a sense—of the characteristic of the sequence of sessions. It is characteristic of the sequence of sessions—not of any particular session—that which was rewarded in a particular session shall not be rewarded in the next. And that is, you see, from going along at this level of intellect to that level—jumping up to an understanding if you will, that the subject matter the—I think one has to use the word subject—is not the single session: the subject which is being discussed is the sequence of sessions. And "that" sequence of sessions has its characteristic. You see, the porpoise could not have done the trick in a single session.

The porpoise has to have the experience of a succession of sessions in order to discover this generality "about" sessions—that no one is like the other. You can't discover that from a single session. So, what the porpoise discovered was presumably non-verbal, so far as the porpoise is concerned—I don't know that—but it's very close to things that we use words for, very close to the business we were talking about at the beginning with Socrates and the class of men who die.

What she has discovered is a difference between the individual item and the class. A very extraordinary discovery, but one which looked at from the backside ... after the fact of the discovery is very simple, you see. But looked at from this side, before you know it, is very difficult to get at. We repeatedly run into "that" sort of difficult—that the discovery, when it's made, is so simple. Why didn't I think of that before? And the process from butting one's head against the insoluble to suddenly seeing the resolution is one of the things that science goes through again and again and again.

And it's one of the things, I would say in passing, that psychiatry and behavioral science, and all that muck has had very few experiences of.... I don't know that I can think of a single moment in the progress of behavioral science that has had this characteristic of suddenly something falling out because of the opposite of what one thought—that one little idea had to be added, and "then" it would make sense. My own double bind is as good an example as I can think of, actually. Well, I think I've occupied enough words. (G. Bateson, 1980)

The porpoise got it. Can we?

I have never figured out what, why, or how some people seem to see this shift I am alluding to throughout this piece. But, I am encouraged by the encounters I have had with people who were able to grasp this difference intuitively. They have been schoolteachers, taxi drivers, farmers, small children, neurodivergent friends, and people I met in other cultures and countless other circumstances through random conversations. There is not any particular experience that readies someone to be attentive to this. It is not easy, as anyone who has been to a school, in a car, or a grocery store has been grooved into thinking in ways that repel this concept. I find myself slipping, even after years of practice.

This habit of confusing categories makes it easy to pop out of a given situation and into an abstraction and then try to solve it. Doctors are caught in identifying symptoms as syndromes and offering treatment at that level instead of addressing the likely underpinnings of poverty, stress, and struggle to succeed in a cut-throat world. Parents are caught pushing their children to excel in scholastic measurements to secure their place in later life. But in scheduling tutoring, extra credit classes, and countless other developmental activities, the children have no time in nature, no time together as a family, and most importantly, no time to be bored—all of which would give their children a sense of membership in life. The research that shows the decrease in insect population is essential, but the response is not to make robot insects to replace the insects' pollination of plants; this is an error of perception logically followed by an error in action. Cultural, intergenerational expectations of desirable suburban yard maintenance, of gardening flowers that local insects don't recognize, of wanting to buy cheap, well-shaped vegetables at the grocery store—all these feed into a demand for extensive agricultural use of pesticides and GMOs—and little robot insects are not going to influence people to desire different vegetables. Likewise, criminal punishment does not teach people not to commit crimes. The context in which the character committed the crime is not changed when the "criminal" is incarcerated. Cries for "law and order" are better seen as cries for help tending to desperate societies.

> In sum, each of these disasters will be found to contain an error in logical typing. In spite of immediate gain at one logical level, the sign is reversed and benefit becomes calamity in some other, larger and longer, context (G. Bateson, 2002, p. 164).

There is a way to respond that comes implicitly with a shift of perception. If you see dog poop on the street, you do not need to send a memo to your foot not to step in it. More than likely, you will step aside without even thinking about it. After years of trying to describe how the practice of noticing differences in abstraction can be helpful to another approach to action, I found myself saying these words:

> *The shape of the response needs to meet, not match, the shape of the trouble*
> *MEET = many connections in motion*
> *MATCH = singular static connection*

To match the shape of the trouble is to strategize only from the trouble as it is expressed and not tend to the undergrowth from which it came or the overgrowth into which a reckless action will further fractionate. You could say this is attending the map. To meet the shape of the trouble is an approach imbued with knowing that multitudes of contextual movements are being expressed—and shifting the possible limits of ecological communication inside the situation. Usually, this looks nothing like that, which correlates to the identified trouble. You could say this is attending the territory.

The unintended damage that occurs across multiple contexts resulting from attempts at direct correctives, solving for the label, fixing the abstraction, and altering the menu fills the local and global news. It is what is filling the hospitals and the prisons; it is what is producing calls for law and order, where care and creativity are needed to meet the many swirling ecologies of possibility. I will try to illustrate the difference through a story.

INTERGENERATIONAL LOSTNESS

My two children and I were visiting a friend one afternoon who lived on the outskirts of Buenos Aires, and on the way home, we got on the wrong bus. Buenos Aires is an enormous city, and I was not at all familiar with the bus system, hardly able to speak Spanish, and it was a Sunday, which meant that the regular schedules were changed and the buses would stop running early. I estimated we were at least an hour and a half from where we were staying in the city. My children were nine and eleven years old. There were no taxis. The moment I realized we were on the wrong bus was one moment too late. We had spent all of our pesos getting on that bus. I was looking out the window at the scenery and landscapes I had noticed on the way to our friend's house. Suddenly, the bus turned, and we were headed somewhere other than where I expected. It was the early days of international mobile phones, so that was not an option either. What to do?

If I had been alone, I might have hopped off the bus, tried to return to where I came from, and started over. But with two kids and a sudden torrential downpour, going back was out of the question. It then dawned on me how important this moment was as a "teachable" moment. Clearly, I could not teach my kids all the ins and outs of the Argentinian bus schedules, nor could I teach them the streets of this suburb of Buenos Aires—but I could be "lost" with them.

What happens when the grown-up is lost and does not know what to do? I recognized that this could be quite frightening to my children. It could also save their lives someday. It was possibly one of the most important few hours of parenting I could have ever given them. We could not match the trouble with a bus schedule and the correct amount of pesos. We needed to meet the trouble with an approach that would be multicontextual and found in the details of experience. How to be lost—this is a great life lesson. Instead of plummeting into my panic, I began to laugh. "Uh oh, we made a big mistake. We are going the wrong way; this is going to be an interesting afternoon." Together, we began to look for familiar landmarks. The kids found a few pesos in their backpacks, and we were able to get on another bus. It was also a "wrong" bus . . . and the rain was creating puddles on the streets almost to the kids' knees. The tension should have been growing. But it was not. We knew that we would end up where we belonged one day, probably later that same day, but in the meantime, we were wildly off course.

The kids were equally as engaged in the project as I was. Noses pressed against the bus window, trying to find something we knew. No one was leading this wayward triad; we were winging it. We ended up in a restaurant where I could pay with a credit card, and a kind cashier allowed us to overcharge the snack we ate to get a little more cash for the next bus. We rode and rode, looking for something to orient us. Finally, we ended up at the zoo. We had been to the zoo before. We found a place where we could speak with a taxi driver. We explained in the chaos of ripping thunder and soggy excitement that we were lost and had no money. He drove us to a bank machine and then to where we were staying. That afternoon, we met people, we went places we would never have visited, we had to engage with people in ways that were uncommon, ask favors, and see where the next bus would take us.

Later, when the kids were in their teens and beginning to go out into the world on their own, we referred to that afternoon as "a way to find a way. . . ." Getting lost in the outskirts of Buenos Aires is, of course, not the same as getting lost somewhere else. The details of that experience are not helpful in another lost moment, yet the attention to detail is. The relationships we made that afternoon are not useful to other cities, but the possibility of making relationships is everywhere. The exact responses in that mishap will never be repeated, but knowing how to be in a mishap and keep a sense of humor is rooted in attention to the particulars of a situation. We did get back, but that was not the solution; that was a consequence of an approach to meeting the trouble. Trouble will come, no doubt of that. It will surprise and upset the momentums of expectation. When it comes, there is no script, no recipe, no formula or code of how to make it go away. No five-step solution will fit the many shapes trouble can find. Improvising response from within its complexity is an art form of paying attention in another way.

What is a Direct Corrective?

More often than not, attempting to fix a problem in a living system is done at the wrong level. The consequence is perceived as the cause. The doctor treats the stress of work-life, the police punish mental illness, and the governments bail out the banks. In each case, the identified problem is located and rooted in contexts other than the one tasked to attend to it. The doctor cannot, in fact, do anything about the stress of poverty, the police cannot treat the depression and anxiety of feeling unsuccessful, and the governments cannot fix the rabid debt culture that the need for increased profit creates. Unsurprisingly, the results in all cases are dismal. Doctors are burning out, social service workers are losing faith in the system, and people are distrusting governments and banks. Causation for these failures is generated from all sorts of epistemological bases. Some will look for political, historical, or religious causation; others will look to technology or conspiracy. In separating the different lines of causality, the attention is fragmented, leading to a multitude of mix-ups in follow-up action plans.

> Brian: "Look, you've got it all wrong. You don't need to follow me. You don't need to follow anybody. You've got to think for yourselves. You're all individuals."
> Crowd: "Yes, we're all individuals!"
> Individual: "I'm not! (Goldstone, 1979, Monty Python's Life of Brian)

The assignment of causation is fragmented, and the issue is cut from its nest of contextual entangling and isolation. The responsibility for health is perceived to be located in the health system, the responsibility for schools is thought to be located in the school system, the responsibility for politics is located in the political system, and so on. However, the offices in charge of playground safety, sidewalk design, or building codes bring people to the urgent care facilities. The street cleaning schedules that clear the ice off the sidewalks in winter become the medical emergencies of broken hips among older people. The children's curiosities are obscured while schools respond to the state's standardized testing regulations. The school systems are woven into many other contexts that are not visible in the curriculum, including the wish parents have for their children to have a good position in the existing system and the need for babysitting to let parents stay in the workforce. The political system is tethered to the many media technologies and the manipulation of common opinions. It is locked in the promise to keep the existing structures that pay for campaigns and create jobs in place and not overturn them. These structures are upheld by history and technology and tie right back into the schools and the health systems. All are wound tightly into economic bondage. So, how is it possible to make a direct corrective?

Again . . . the shape of the response needs to meet, not match, the shape of the trouble

The meal meets the needs for emotional, nutritional, and communal nourishment—it does not match. To match would be three different tasks. The elegance of how nature meets interspecies ecology—meets evolution—meets ongoing life is an inspiration toward another approach that speaks of grace.

PROMISES THAT MEET, NOT MATCH

The world is full of mistakes in the perception of abstraction, from the most intimate relationships to the global. At times, the intimate realm makes it more difficult to perceive because it is hard not to get caught up in daily drama when we are close. But it can also make it easier to perceive the thickness of stories residing in the situation. These can be dire, violent, and justified by their own logic. They beget destruction, humiliation, and separation. Most importantly, these errors in perception lead to the pursuit of questions, strategies, and policies that are further distractions from the tending that is needed. Measuring a child's math abilities with the use of a standardized test is an illustration of a fundamental misunderstanding of children, math, and tests. Children learn all day, not just in the classroom, and math is everywhere, especially where it is not relegated to numbers but instead lives in rhythms and patterns, which all children are sensitive to in different ways. Tests do not measure the child. They measure the structure of the system that relies on the test itself. I do not want to abandon my child's curiosity in favor of a societal error.

The one-size-fits-all models of development, social well-being, health, or success—or even of ecological, communal living—are violent misunderstandings of the contextual possibilities and diversities. No matter that they are designed in hopes of achieving efficient programs to keep people in the workforce—they inherently disrupt the necessary complexity of life. There are 2nd-order and nth-order consequences to this violence which emerge in other contextual ways: political rebellion, sickness, mental health issues, robbery, racism, and so much more.

To describe a living system—a child, a family, a forest, a lover—without killing complexity is to enter the descriptive process with attention to how the living system can continue to change whilst being in the description. Does the description stop time? Does it cut contexts? Does it allow unpredictability?

When we were thinking about getting married, Mats (my husband) and I began by asking one another quite seriously—what do you want me to promise you? We have a unique story of having met early in our twenties in Thailand, having had a brief love affair, and then losing touch for twenty-five years before we found our way back together. I think we both knew over the years that our early romance was more significant than we could admit at the time. Neither of us ever thought we would see the other again. In the twenty-five years that passed, we found other partners, had children, and built our lives. Our respective marriages both ended. We found each other, fell back in love, and wanted to claim this second chance. This time, we wanted to be careful not to cause the kind of hurt we had experienced in previous partnerships. Both of us had been through divorce. Both of us knew of the painful grooves that couples can fall into. We did not want to put each other in that pain. We were in our forties then.

Marriage, for us, was not about starting a family; we both had children already. For us, this was a union that required attention to the ways in which we could get caught in cultural scripts that limited who we could be. These scripts were engraved in the usual vows. Promises to be together always. Promises not to fail the marriage . . . but as much as I love Mats and he loves me, I do not want him to be with me if he does not want to be with me. That sounds like hell for both of us and our kids. We considered this carefully and came up with our own ceremony and vows that included how we wanted to meet this idea of a commitment or a promise. This is the ceremony we had.

Vows and Ceremony

Officiant reads:
This is not a traditional wedding, but tradition is with us. We are not traditional people. It would be foolish to apply the usual vows to the union of Mats and Nora. They will still say, "I do" at the end. Don't worry.

It is not necessary for Mats and Nora to promise to love each other always. This is true because they already have always loved each other. They do not need to say the words "till death do us part"—they have already been apart and have come back together.

This marriage marks the continuation of their togetherness through time and experience across decades, generations, and cultures. They are catching the same warm Thai breeze that blew them together so long ago.

This is not a beginning. This is their infinity—the bond they share is not measured in years or in structures, it just is, and they just are—as always, for always . . . on their way to the essence of themselves.

We are not in a church but are gathered instead in this ecology where our witness is nature. The waves, the sky, and the trees can seal this union. One illustration of nature and of this marriage is music. Bringing these families together is an improvisational symphony.

As the stories of the Anderssons, the Qwarfordts, the Batesons, and the Brubecks come together, they overlap their stories, their patterns, and their rhythms. They are in a constant process of creating this music together. Every tone counts; every chord and, beat, and melody that arises becomes their song. Or our song . . .

The intricate lacing of these worlds is beautiful and strong. Together, all of us have many perspectives to draw wisdom from. We move less blindly with the many visions we have between us. Ours is the music of caring, curiosity, and adventure. This particular family is pretty freaky. We are a strange bunch of characters; all of us are a bit weird, and we are an unusually loving, messy, funny company—free from the expectations of others. This marriage is based upon these promises.

Mats and Nora repeated these words to each other:
I will promise you these things. . . .
~ I promise to show you my whole self in so much as I can.
~ I promise to speak and be in truth with you. I will not hide anything.
~ I promise to learn with you and from you. I will share incomplete ideas and unknown feelings with you so that you know where I am, even when I am lost. I choose you.

Officiant reads:
Mats and Nora, you have offered your promises, and they will be reflected in the trust you share. You trust each other already, in body, heart, and soul. But you share your lives with all of us here, which makes this trust larger than the two of you.

Mats and Nora, you can trust each other to be there in support and love so that you can put your children first when they need you. You can trust each other to bend with your cross-continental life. Certainly, this family will stretch across geographies and cultures together. Everywhere will be home, and nowhere will be final.

The officiant asks Mats:
Mats, we have said that your union is an improvisational symphony. You do not know what will be required of you both in the years to come. Do you trust Nora to choose your path with you, together in integrity and love, as life carries you along?
Mats: *I do.*

Officiant asks Nora:
Nora, we have said that your union is an improvisational symphony. You do not know what will be required of you both in the years to come. Do you trust Mats to choose your path with you, together in integrity and love, as life carries you along?

Nora: *I do.*

Kiss.

Time changes us as we twist into responses to each other, our past, our hurts, and the wiggles in our life paths. To endure togetherness through time is not to stay the same but rather to continue to mutually compose our life songs. There is the chance that we may grow apart, that our worlds will configure as un-congealing, un-mixable potions. That is the risk of being alive. The alternative is a promise of stagnation, which is worse. The seduction of control leads to distortion, contorting us into shrunken versions of ourselves and our union. The love that is shared will not be just one sort of love; it will be many, and they will change as we do.

Meet me, don't try to match me. I will meet you in many ways—never the same. I would want you to change, want you to learn, want you to find life interesting and wonderful in new ways as we go through time together. A map "is not" you. You are impermanent. As am I. Our details will form entire landscapes of possibility. So the promise I make, if my promise is in the grace of our movements through time and experience, will be able to hold the "you" that does not yet exist and the "you" that once was. Our togetherness is more than our promise to one another; it is a promise to live. Promise me you will keep going.

> Do ideas really occur in chains, or is the lineal structure imposed on them by scholars and philosophers? How is the world of logic, which eschews "circular argument" related to a world in which circular trains of causation are the rule rather than the exception? (G. Bateson, 2002, p. 18)

```
We are in the tumble of the wave
      And it is blurry ...
           dizzying ...
      It's tilted and grainy.
```

Fig. 4. Nora Bateson. (2023). *More Than Blobs*. [oil on canvas]

HALLWAY OF HALLWAYS

Trying to fix the wrong context of broken is a hallway of hallways.

The suffering is the consequence, not the problem. Don't solve it.

It's not that the schools are not good enough - it is that the next generation is being squeezed into the past.

It is not that poverty is devastating - it is that the economy devastates.

It is not that our bodies have been violated - it is that predators are celebrated.

It is not that the tech is evil - it is that it is unable to feel.

It is not that the doctor is wrong - it is that the poison in food, water, air, and stress are beyond the reach of medication.

It is not that politicians don't make change - it is that they are bound to keep things as they are.

It is not that the journalism is crap - it is that it is selling one side of a 1000-sided story.

It is not that the climate has problems - it is that identity is mixed up with material wealth.

It is not that there are dangers out there - it is that people need each other and are taught to hold back.

No committee or action team can actually reach the conditions that produce the issues.

But they can produce metrics on solving the consequences.

It is not that there is no wish to meet the deeper needs - it is that healing cannot be divided or measured by department.

The confusion is compounding.

Going down a path that starts with an inadequate task is the beginning of an assembly line of industrial-sized overlapping systemic problems.

MOVING EDGES

Living is moving always ...
Moving among living moving learning forms always forming.
Responding.
Even in stillness,
Even in quiet,
Even very slowly.

Moving is affect ...
Each shift propelling, quickening, stimulating,
Each pressure arousing, bending, reaching,
Each sound resonating, beating, humming.
Each communion rending, touching, stirring.
Each metabolism energizing, appetizing, tissuing.

Even in death, there is movement.
As I change, my memories are in new territories.
My relationships with those who have gone continue to move.

Algae bloom in the stagnant water.
The decomposition is nourishing other life.

Static is the freeze frame, the snapshot held against a sky of moving clouds. This is the illusion of the map, an abstracted idea of something that is outside life and somehow untouched by its ever-moving-ness.

Stuck-ness is not a stop ...
There is wild compensatory movement around it.
Change is everywhere all the time, this is ancient knowing.

It takes a lot of change to keep something stuck. Draw your circle around the stuck bit and then zoom out to see what it takes to brace one piece of a moving circus against all the contexts in play.

The eternal is not framed ...
Always reframing from infinite directions.

The sacred is in "the motion" itself.
The infinite is in constant reshuffle.
An always of morphing swirls.

And when I am moved, I am emotive.
Meaning is squishing into new insights, poignant ...
Ideas are shaping and twisting against each other.
Warm rushes of being in the moving world, emotions in motion.

Healing is movement, learning is movement,
moving the moving into new movements.

The ripping and the tearing is also the movement. Destroying.

It is like entering the Double Dutch jump ropes,
Catching the tripling rhythms in poly-pitter-patter ...
... Step, hop, swish, hop, step, jump ...
... The beat is moving.

Deep thuds, now rain-like, now hearts-in-sync, now breath.

Find the entry, feel it in ...

The information is alive.

The response must also be alive.

This is a movement.

MAMA NOW
(For my children - What it's like to be your mama)

Your eyes will see the derailing of assumptions,
Your hands will hold the crumble of the old matrix.

I do not have any authority to lean into,
I have empty pockets where parents used to advise their children,
I do not have any maps, myths, or mother-wisdom for you.

I can fix your breakfast, but not the culture,
And when you ask about how to be a good person,
I cannot lie to you.
Everything you touch in a day is in some way bloodied.
You have been born into an edgeless violence.
But I will not judge or measure you against a bygone metric.

I am here, too, ready to learn with you.
Unsure how to be or who to be.

I can only read fragments of your worry,
As the future is a horizon of confusion.
I cannot protect you. And yet it is my only job.
Aching as I witness from this side of the hourglass.

Other generations of parents knew the outlines,
School, career, family, and retirement.
But your life will be another shape entirely.
Forming in the fractures.

When you say you need a goal, I offer you an expired ticket.
Superficial memes roll off the tongue right into your bullshit detector.
Success in the existing system is not going to do you much good.

Your integrity is your rage, and I will nourish it.
Your dignity is your curiosity, and I am tiny beside it.
Your courage is your pain, and I will sing to it with you.

We will riot together.
We will notice the nuance of small graces in the day.
We will wash the grit of loss for each other.

I am your mama, and your future is the story of a storm.
I am your cabin, your boots, your rucksack.

JUICY

But isn't it interesting that the erotic is so
removed from the study of life? How?

And how?

And how to bring it in?

There has always been a school of lovers,
rigorous in their studies –
They could not-not see, could not-not describe life
as juicy.

But the bookishness held its sway on slabs of
concrete and stainless counters.

There there is no place for lusty mossy forest
floors and shameless blossoms to explain themselves.

The real flat-earthers are the ones who make life
into a flat description.

It will not serve now.

Now we go into the breath,

So we can breathe.

Now we go into the sensual,
So we can sense.
Now we go into the intimacy,
So we are not alone,
Now we go in.
Or we go out.

THE CARAMELS OF AUTUMN

Butterscotching birch trees,
And honeying maples,
In their confection perfection,
For a few soft taffy hours,
They cherry in candied red at the forest edges,
The landscape is sugar-shined in dampness,
Tilled wet earth as heavy as acres of chunky melted
chocolate, waits —
Apple-green grass glows in sour snappy light,
While cola-bark oak trunks and branches gum and
crunch,
Milky Earl Grey sky pours from porcelain,
So Saturday sweet.

Combining

> "The nourishment of Cezanne's awkward apples is
> in the tenderness and alertness they awaken inside us."
> ~ Jane Hirshfield

Un-pick-apart-able*

A small, scratched hand is pulling ripe, red currants from a bush, plucking generational knowing, while the child is chitter-chattering in the warm evening air, gritting the nutty seeds between molars, brushing an ant off a sunburnt elbow. The bucket is full to the brim with berries the neighbors did not want. Tomorrow morning, these berries will go on yogurt.

This piece is not about the importance of eating organic food, local farming, and slow-food recipes. Nor is it about the damage done in the name of consumer culture. I am assuming that, by now, those topics are given and that if you are reading this, you have noticed that there is a distinct possibility that no matter how much we recycle and buy organic, there is a good chance that the human species may not make it through the coming climate changes. You may have also noticed that cultural/political fissures may kill us first. You may have children in your life and find yourself longing for a magical time bridge to deliver them safely to the other side of the heavy shakedowns of the coming years. The instinct to provide a world of future life and love is strong. But is it more potent than the idea that tomorrow will be like today? Ironically, the possibility of continuing generations requires discontinuing current ways of living.

There is salt ahead. Stinging and necessary.
Staving off numbness and the scurvy of disconnectedness.
There are whole grains, fermented ideas, intoxicating and nostalgic.
The past is as lovely as it is deadly. Take small sips.
Add logic and enchantment to taste.

In the everyday gesture of a parent providing breakfast for a child, the entire future of humanity and thousands of other organisms pivots. Like other animals, humans are tasked with feeding the next generations. Life is dependent upon this seemingly simple mandate of continuance. Feed the babies. Don't fail. Bringing a morsel of food to your lips or the lips of another is an act of intimacy. It is a personal contact point with the seasons and the generations. The tiny act of sharing a meal connects our intestines to the rainfall and the strength in our muscles to the recipes of grandparents. The weaving of baskets to carry the harvest is sewn into the engineering of farm tools, as well as the patience for and knowing of the land. Food is not made of nutritional measurements; it is made of intimate relationships. My sister once said, "Human beings do not eat nutrients; they eat food" (M.C. Bateson, 1994). It is in this intimacy that lies a possibility for system change.

Combining

The big picture—world-saving, global activism, and so on—all of that is important. However, the conceptual shift that makes substantive cultural change possible is the ability to perceive the vital complexity of each interaction in a day. That shift is accessed through the most personal unions with the world around us. Sex is one such union, with the complexity of our bodies coming together in communication. Like sex, food is a medium through which life in the external world becomes internal communing, for better and for worse.

> *It is what was "normal." It smelled like home. Loved and broken.*
> *The cinnamon of expectation, family bickering in the kitchen.*
> *The bark of trees, the barking dog,*
> *The perfect table, the gathering, the tarragon cadence of relatives,*
> *Halfway connecting to each other through the feast and retelling stories,*
> *Halfway disconnecting in slicing comments and garnishes of dominance.*
> *Peppered in debt, not just to credit cards,*
> *Debt against ourselves.*
> *The flatware clinks.*
> *The elders sigh and do not say things that need saying.*
> *Still stewing.*

The contexts of personal history, media, genes, and religion are just a few of the many influences that make people prefer particular flavors, temperatures, and times of day. The sauce of that meaning-making is aged and un-pick-apart-able. When non-verbal realms of the body interface with culture, politics, economy, medicine, the arts, and history, there is a transcontextual mingling of meanings—all in a single meal. Of course, it is impossible to achieve clarity on what meanings are made in that murky mixing of territories. But, in lieu of clarity, caution, and care toward more contextual sensitivity, it is a big step toward living in a new way. And it is clear that the whole mess, from memories and textures to conversations and emotions, becomes integral to well-being.

Emergence is the term that captures this complex vitality of blurry relational possibility. It is what happens when countless streams of interaction come together. Emergence is a treasured source of possibility in this era, gilded by the new age and imported into organizational management as the next "it" thing. But *emergency* is also emergence.

Food security is an emergency. The inability to feed babies for any reason is an emergency. And, in emergencies, it becomes dramatically challenging to keep the romance for emergence alive. The lens of perceived possibilities slams shut and starts spinning out binaries of them and us, this or that, live or die. In crisis moments, the perception of relational and interrelational processes becomes seemingly gratuitous, and only the practical is given legitimacy.

In this era of ecological and cultural upheaval, it is becoming routine for government agencies and humanitarian organizations to anticipate big floods, earthquakes, wars, droughts, or fires, interrupting supply chains and agricultural production. Experts can quickly translate the need to feed starving people into the rapid delivery of a million food pods with calculated nutritional benefits. The rest of the interdependencies vaporize under the justification of addressing urgency. But short-circuiting complex systems is never the best plan. The consequences, and consequences of consequences, will continue to make themselves known. Again, food is made of relationships. Providing food, even in emergencies, supports the relationships through which food emerges. Feeding the babies is an act of love that is as necessary for children's parents or guardians as it is for the children. The nourishment of food and water is imperative, and right behind that is the relational field of nourishment.

Communing and Consuming

Every morning worldwide, billions of breakfasts are prepared, served, and shared. There is so much for breakfast: pickled fish, or boxed cereal and toast, or rice and pork, or cappuccino and pastries, or eggs, or avocados, or noodle soups, or smoothies, or chia porridge. People with sleepy eyes open refrigerators each morning. They put on tea kettles and swallow up their meal. In doing so, the interdependent systems of our world are extended into another day. But there is a double bind at the breakfast table. This extension of business as usual into each day is stepping ever closer to environmental, cultural, and economic suicide.

On the one hand, to survive, we must feed our children breakfast. On the other hand, to feed our children, it is necessary to participate in inherently deadly systems. We will have to pull back from all forms of exploitation to protect the possibility of breakfast for the babies. In that statement is the imperative for clean oceans, for gender equality, for protection of the forests, for human rights, and to end both poverty and wealth.

What will it take to provide a world where it is possible to feed the babies, where the soil is teeming with microorganisms that meet happily with their associates in the intestines of breakfast eaters, and where the people who harvested the grains can also feed their babies? It will take failure and betrayal in equal measure—blended until smooth, then poured over the top of marinated remorse and baked at high temperatures until the glamorous memories of this era of having-ness melt in your mouth and the assumptions of global economies are digested by time.

I was stunned recently to learn the etymology of the word "consume."

> consume (v.)
> Late 14c., "to destroy by separating into parts which cannot be reunited, as by burning or eating," hence "destroy the substance of, annihilate," from Old French consumer "to consume" (12c.) and directly from Latin consumere "to use up, eat, waste." (Etymonline, n.d.)

I was surprised that the term "consume" was not more oriented around swallowing, taking in, or absorbing. But, when I learned of this etymology, I could not help but think, "No wonder."

Supporting the Support: Warm Data

The way of thinking, perceiving, and conceptualizing that "consumes"—breaks the world into pieces, separating the necessary connections—provides the illusion of objectification. To perceive any aspect of a living system as objectified is a dangerous business that rapidly becomes exploitation. And here we are in a tapestry of cut connections. What, then, might the opposite of "consume" be? I might have suggested fasting, "getting rid of," or a form of austerity. But the definition, "destruction by separating into parts," is all too appropriate, so the opposite must be something like "reassembling toward life."

The question, then, at the breakfast table is how to support the relationships that together provide the morning meal—how to support the support of food. I need a strong coffee before playing with 2nd-order questions like that. It is possible to begin to imagine some of the contextual processes of which food is a consequence, starting with describing food as relational.

To do this, one must notice the warm data—the contextual and relational information. It is not enough to simply list the various sectors of socio-economic structure that are needed to provide food and stream them together in hopes of gaining an understanding of context. That information may be productive for some projects, but to get at the warm data is to look for the relationships between the contexts—the connective tissue. This kind of information is often not measurable or repeatable and is undoubtedly not "objective." The observer matters.

Warm Data Toggle

The relational information begins to zoom in and out between these few contexts.

> *The glorious lavishing affection of life's bounty spread across a table in steaming dishes, flavors, and dimensions of farm-to-table vegetables, succulent fruits, aged cheeses, fresh breads, and smooth creams.*

> *Eight hundred million people will go to bed hungry tonight.*

> *The lettuce in the supermarket across the street from where I live in Stockholm has never seen sunshine or soil.*

> *As the climate changes, sixth-generation farmers in Spain who grow tomatoes will be unable to continue their crops. Should they move to new land or new jobs?*

> *Food-related illnesses, including diabetes, obesity, eating disorders, and others, are rising globally at alarming rates.*

> *The recent heat waves left livestock owners struggling to feed their animals. Suddenly, there was a lot of meat to sell.*

> *My children saw the movie "Holes," in which onions were eaten by the kids to protect them from poison lizards. They love onions to this day.*

Transcontextual breakfast is made up of the relationship between the generations and the culture, the soil, the bacteria, the family, the agriculture, the economy, the market, the body, history, technology, and the bodies of the diners. There are many more contexts. There are always more. The way in which these contexts resonate with each other is what is interesting. Food is there, in the in-between, in the relations. To imagine a future world in which there is breakfast for the babies is to imagine that these relational processes are supported—or, at the very least, not disrupted. This is not a sterile project. Supporting and building relational health within living systems requires peripheral vision, indirect actions, and wide-angle observation that catches not only the result but also the rhythm, emotion, and warmth. It is called warm data because it is information that has its relational nest included.

Food pods fortified with vitamins and minerals are what happens when the relational nature of food is not noticed. They may be necessary to stave off death, but caution is advised around what sorts of consequences will "emerge." Just that sort of short-cut-severed-from complexity thinking got us here, to begin with. There is damage to the fabric of the community when mothers and fathers do not dish out warm food to their children. Even in the eye of the storm, it is an excellent habit to notice and respect the way food knits together a community. Modern living has done vast harm to these processes and relationships, but they are not gone, only reorganized. Look carefully. There is poetry in breakfast. It is a vital ingredient.

Breakfast is the agriculture of growing grains and fruits. It is the industry of meat and processed foods. The distribution and sales of food products make up vast economic systems that reach around the globe. The pesticides used to grow the grains are ruining the soil and forming an industry of chemical research, development, marketing, and sales. The labor violations of coffee and tea plantations are poured into cozy morning wake-up cups. The fish that is served carries the toxicity of modern oceans. The tankers, trucks, and factories that produce and package our morning sustenance are degrading the atmosphere. The animals that provide the dairy and meat are pumped with hormones and treated with disgusting cruelty. My clothing is made by women knee-deep in carcinogenic dye for sixteen hours daily. Everything my eye falls upon in my living room is made possible by exploitation. I cannot put any of it back.

I cannot fix it. But I had better not deny it. No matter the difficulty, I welcome the tangled mess and bring it forward into my perception. Let me grasp it with as much contextual and transcontextual description as I can muster. To welcome the complexity now is to meet it, to know it is more significant than initially anticipated and coming faster than the experts predicted.

When the emergencies emerge, and they will—they are already coming—I hope I will remember that the relational response is the one that is harder, more expensive, and takes longer. I will not shy away from the trouble it takes to heal from the inside out. Responding to food emergencies with emergence requires attending to the many relational processes that result in a future in which babies can be fed. That is the opposite of the word "consume."

*Previously published in *Kosmos Journal*

Fig. 5. Vivien Leung. (2023). *Blooming*. [digital art].

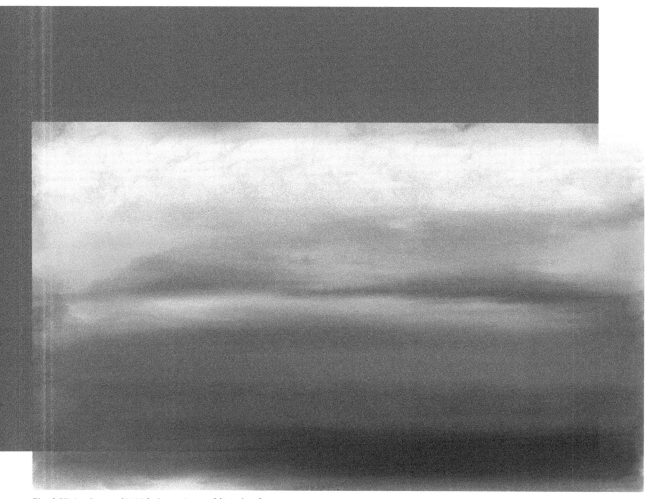

Fig. 6. Vivien Leung. (2023). *Ocean Breeze.* [digital art]

COMMUNING

Instead of silence,
We speak in situations.
The language moves into the river.
Below the hunter's traps,
Where the rabbit warns the deer.

Giggling in the secret code ... like siblings.
There is nothing in the flat speech to say now.

But there is everything to commune.

UNCUT

Every note lands in a teapot of other notes ...
What are they brewing?
Is it in integrity?

Those relationships are making more relationships –
What have I contributed?
How is the ring of my being singing through others?

Have I shown up with my many selves?
What did I perceive I need to edit to be in communication with you?
What did you need to edit to be in relationship with me?
What did I cut?
What has been trimmed, so that I could fit you into my illusions?

What will the wounds say?
How will they brew with time?

Are we teaching each other to hold back?
Your rage, and your wonder, your scratching at life's scabs,
And still your soothing fingertips' grace.
Both are consequences of time across generations navigating life.
Talking about a future, don't stain it with any utopia,
Terrorizing the necessary complexity.

Would it help if I said we are learning together?
Shaping each other.
Unsliced, integrating, reunion.
To be uncut with you.
Uncut in our uncut world.

Combining

A Pineapple Surrounded by Cockroaches

There was a time when botanical drawings always depicted multiple organisms. These paintings from the 1600s & 1700s are so telling. The interrelational ecology is unquestioned. Later botanical illustrations depicted single organisms. Then parts of organisms.

Fig. 7. Maria Sibylla Merian. (1701-1705). *"A pineapple surrounded by cockroaches."* [Watercolor and bodycolor on vellum]. Image from the British Museum. London, United Kingdom.

Tarantulas, Hummingbird, Spiders and Ants

Science and art are combining here to give us relational information. The zoom in and the zoom out. Reductionism, and context, together.

Botanical artist Maria Sibylla Merian was born in 1647.

Fig. 8. Dorothea Graff or Johanna Herolt after Maria Sibylla Merian. (1701-1705). *"Tarantulas, one attacking a hummingbird, spiders and ants, on a guava tree."* [Image of Watercolor and bodycolor on vellum]. Image from the British Museum. London, United Kingdom

COMBINING

"How can I begin anything new with all of yesterday in me?"
~ LEONARD COHEN, "BEAUTIFUL LOSERS"

Where is The Edge of Me?

One of the difficulties of being so certain of the change one thinks one wants to see in the world is that it is challenging to discern where the external contexts end, and our identities or selves begin.

Think of the snowy owl, like the moths, whose physiology is reflected in the context it has lived within for so many generations. To live within the structures of the modern world and to find some semblance of health and sanity in it is to have an idea of who I am in relation to my contexts of life. Given that the patterns of life I need to survive are exploitative and extractive, some hard questions arise. I am like the snowy owl, catching my sense of self from the world around me. I am soaked through. I am a crooked tree.

When one says: "That person is compassionate."
"This person is honest."
"The other person is irresponsible."
When one says: "That child is violent, that teacher is empathetic..."

When one uses these terms, one must remember compassion, honesty, irresponsibility, violence, empathy, and all of those descriptions exist in the relationships, in the context—and only partially in the mysteriously combined creature sometimes referred to as the "individual."

People are compassionate in one context, violent in another, responsible in one context, reckless in another, honest in one context, and manipulative in another—the character description describes the character of the relationships, not necessarily the individuals.

This does not mean there is no way to learn; it means that the learning is mutual and transcontextual.

When I am with some friends, I am more serious; they probably think I do not have much of a sense of humor, but I am more witty or silly with others. With some people, I am suddenly able to be wise, while with others, I am baffled. I reveal varying depths of my private life as is viable with different people. I withhold and edit myself, I improvise myself, and I am reflected through the "me" I can be with you in the wide spectrum of my encounters. I grew up around all sorts of people who sold promises of methodologies on how to make people better, more developed, more compassionate, or rich, or righteous, or whatever. I was in the belly of the human-potential movement beast in Northern

California in the 70s. People marketed self-help programs by the dozens, or spiritual practices with wealthy gurus, or psychology on tap, all preying on the notion that "you should be at a higher level of development." My father would raise one eyebrow and flip the epistemology of this thinking like a pancake. I was young then. I did not understand at the time how he was advocating for the complexity of each of us.

Later in my own work, I noticed a strong aversion to defining the "best outcomes" or "what you will get out of this work." I was very sensitive to projects selling things like "courage," "trust," or other identified character improvements. I have gotten myself in hot water protesting this attention to individual development. Where is the edge of me? Where is the edge of you? Where is the edge of the breakfast cereal? The bacteria in the soil that grew the grains? The decomposition of the plant and animal life that feed the bacteria? Where is the edge when I am mostly made of non-human cells?

When I ask, "How can I be a better person?"—the question carries an illusion that is out of kilter with families, cultures, and histories that we are all responding to.

I would rather ask, "Who can you be when you are with me?" This is a more ecological question.

I do not want to be part of any project that has at its core the idea that people can be told how to live, think, or feel. I am much more interested in situations and relationships that allow people to express themselves in new ways. Then, let's see what happens at nth-order. How do those conversations and interactions inform other conversations, and how do they shift the tone of other interactions?

I am leaning into my father's work here. Keeping this piece close at hand:

> In describing individual human beings, both the scientist and the layman commonly resort to adjectives descriptive of "character." It is said that Mr. Jones is dependent, hostile, fey, finicky, anxious, exhibitionistic, narcissistic, passive, competitive, energetic, bold, cowardly, fatalistic, humorous, playful, canny, optimistic, perfectionist, careless, careful, casual, etc. In the light of what has already been said, the reader will be able to assign all these adjectives to their appropriate logical type.
>
> ... In the punctuation of human interaction, the critical reader will have observed that the adjectives above which purport to describe individual character are really not strictly applicable to the individual but rather describe *transactions* between the individual and his material and human environment. No man is "resourceful" or "dependent" or "fatalistic" in a vacuum. His characteristic, whatever it be, is not his but is rather a characteristic of what goes on between him and something (or somebody) else.
>
> This being so, it is natural to look into what goes on between people, there to

> find contexts of Learning I which are likely to lend their shape to processes of Learning II. In such systems, involving two or more persons, where most of the important events are postures, actions, or utterances of the living creatures, we note immediately that the stream of events is commonly punctuated into contexts of learning by a tacit agreement between the persons regarding the nature of their relationship—or by context markers and tacit agreement that these context markers shall "mean" the same for both parties. It is instructive to attempt analysis of an ongoing interchange between A and B. (G. Bateson, 2000, p. 303)

So, in that case, what is system change? What might it look like to allow the world to change in such a way that we don't recognize ourselves in it? . . . And what might it mean for that to be a world of vitality, healing, and mutual care? How might perception of self in context be shifted? Would this allow all the relationships in the physical, emotional, and intellectual realms of day-to-day life to reveal the needed shifts?

Fig. 9. Rachel Hentsch. (2023). *Assholery*. [digital art].

Fig. 10. Vivien Leung. (2023). *Combining*. [digital art].

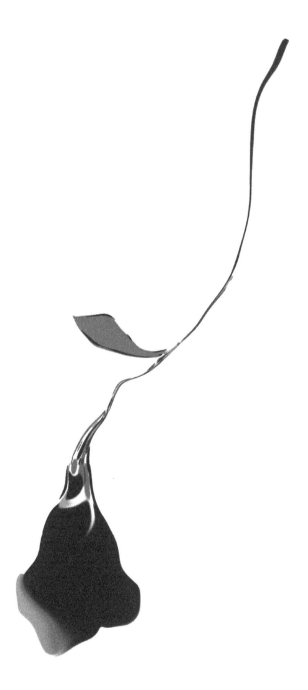

Fig. 11. Vivien Leung. (2023). *Bellflower*. [digital art].

ONE
THING
I
will
share
with
you
is
that
I
don't
ever
want
or
need
to
change
you
or
anyone
else.
Your
changing
is
yours,
&
is
particular
to
you
in
your
complexity.
I will tend the tones, the textures, the implications of how we are in relationship.
Opening
unimagined
Possibility
...

COMBINING

WITHOUT SHIELDS (THE VOICE OF CHANGE IS CHANGING)

There are ancestries and ecologies that are speaking. I am honored when their voices run through me. Only when my integrity is clean is the frequency audible.

Simplicity is complexity with grace.

To be their vessel: to hold the nourishment, to wear the breath of any possible future ... is to cast aside the costumes and scripts of excuses for the damage.

The exploitation that has been justified has bled through now. The language, the status, and the authority once wielded to make the vulnerable quiver now makes cuckolds of anyone who would stand in for the way things have been.

Time's up.

So I stand naked, in the fire, alone in the dark night. Warrior-ready to simply disavow the traps of materialism.

Eyes rolling in disbelief. Once again, the presumption ... the nerve is remarkable. Taking, tricking, claiming is the perverse providence of anyone who would adopt the stance of the oppressor. The sociopathic eagerness for wealth, influence, at any cost, is such a small shoe. No voice of life will speak through that craving mess.

I see now that this work is not daytime.

I am walking barefoot through the glass of broken worlds.

I have been training for generations for this. Do not try to shelter me, my soft tissues hold the fluid of forever.

I can hold this pain with joy and wash each day with the wholeness of great-great-grandchildren who will one day play here.

May they have soil under their fingernails.

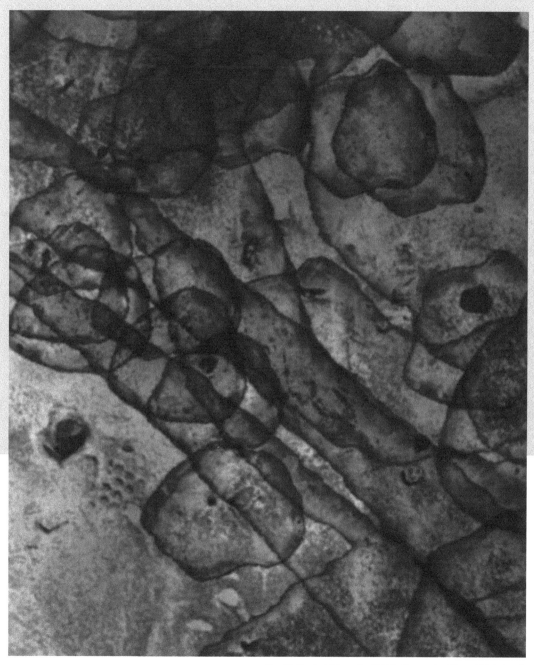

Fig. 12. Rachel Hentsch. (2023). *Connective Tissue*. [digital art].

Combining

Fig. 13. Rachel Hentsch. (2022). *Expressing.* [digital art].

Finding a Way

Will People's Responses to the Emergencies of the Coming Decades Be Warm? Or Cold?

The crises of the moment do not need further description here. Suffice to say that the complexity of the overlapping crises of inequality, health, justice, technology, ecology, and culture are producing emergencies that the institutions of the last centuries cannot contend with. How will the next decades play out amidst these crises? More importantly, what is possible for societies around the world to learn in the process?

We, Nora Bateson & Mamphela Ramphele*, are bringing our voices together on this topic to open a conversation with a conversation. We have both been watching these patterns for decades, from different sides of the world, different experiences of life, different cultures, different generations... but we have seen something very similar. Each of us has struggled to find the thing people call "traction" for these observations we share. We have been told it was not practical, strategic, or policy-oriented. Today, we are writing this together, mutually learning how to express this fundamental shift in approaching environmental and social change. It turns out it is not traction that is needed but relationship.

The lifeboat story, the cold version: Garrett Hardin (1974) introduces an old metaphor in which a lifeboat is featured as a way to ponder the mathematics of survival. The thought experiment is explored through the lens of the ethical distribution of resources. It has been applied to overpopulation, immigration, natural resources, food supplies, and more. High school students are often given this "lifeboat ethic" as an introduction to how hard decisions at structural levels are assessed.

In the story, fifty people are in a lifeboat that can hold ten more people, but there are one hundred people in the water. It appears that approximately ninety people would be left to die so that any of them may survive. The project then becomes a collection of "difficult but necessary" questions to justify how the remaining ten people will be selected to survive.

Who will choose who lives and who dies?
What are the criteria for those choices?
How will the ones who are on the boats be fed?
Is cannibalism a possibility? (Really, this is one of the questions.)

These. Are. Cold. Questions.

COMBINING

The above questions are laced with the thinking of eugenics. They are carrying the darkness of a world in which some people's lives are measured with icy objectification and some are... justified as disposable. They are questions that provide us with "information" about the conditions of the prevailing systems that create the conditions for those sorts of questions to be asked, considered appropriate, and acted upon. And right now, those conditions are cracking.

This sort of approach begets cold lines of reasoning and responding. Perhaps it is an extension of the same coldness that has justified the exploitation and extraction of people and natural resources over the last several centuries. Willingness to objectify by assigning numeric metrics has allowed the illusion of control and management to infect even valiant environmental and social activism efforts. Is it possible to respond with a warmer approach?

In another version of the lifeboat story, like the first, we start with fifty people on the boat, a hundred in the water, and ten more seats. This time, the people on the boat start figuring out how to make it work. They improvise, learn, and tap into the complexity of themselves and each other as a source of unlimited possibilities. In the Warm Data work, this is referred to as "life-boating"—in direct contrast to Hardin's version of the story.

People are not numbers. Those hundred people in the water are not bobbing numerals; they are human beings with histories, experiences, cultures, and languages. They are complex. Numbers are just numbers; they cannot source possibilities from each other or find a way when the basic arithmetic says there is none—people can. Abstracting the solutions to numbers inherently dehumanizes and unnecessarily constrains the spectrum of possibilities. The metric logic removes the human breadth of experience and relationship. This cold calculation flattens the scope of the thinking. We know that if one of us were in the water with ninety-nine other people, each of us would find a way. This is the magic of being a living organism. The quote from the film Jurassic Park says it clearly: "Life finds a way."

People are not roles—a word of warning: collaboration can quickly become a mechanistic allocation of effort according to roles. Alternatively, as it is in life, the living response is one of incalculable and often surprising alternatives that the reduction of the original question to simple arithmetic completely obscured. The quantitative question eliminated the multiplicity of contexts, while a warm approach invites them back into the relationships. This is not a matter of assessment of who is good at what and assigning roles according to expertise. On the contrary, "finding a way" in this situation is about the unique possibilities that occur in the relationship between particular people in that particular water on that specific day. There is no formula, no method; this realm of possibility is accessed through a sentiment of human care and imagination.

Perhaps some people would take turns swimming; maybe others would tie clothing together to pull people; possibly some would hold others' hands and pull them. Each grouping of people would, of course, have a different, particular set of characters and histories that provide their own potential responses. We are all capable of different things depending on who we are with and what they are finding themselves capable of. Capacity cannot be front-loaded; it is emergent. But it is possible to front-load a baseline perception of self, others, and the world that assumes the inherent multitude of stories and draws from them. With care and imagination, the possibilities are endless.

Particular complexities find a way... the possibilities for finding a way in emergencies are not countable, controllable, or predictable. The way in which it becomes possible to respond depends upon the unique lives of the people; it depends on the water that day, the clothes they are wearing, the food they have eaten, and the experiences in their lives that they had prior to being in the water.

The solutions to the situation are entirely contingent upon detail—they are accessed spontaneously, iteratively, and simultaneously. They are as infinite as the combinations of people.

The cold questions are distracting. Possibility in the warm lifeboat story depends upon the likelihood that people will see each other in solidarity to find a way. If they ask questions about the cold numbers, they will be worried about who gets to live and who is chosen to die. In life, this looks like polarizing societies with each group vying for position in the distribution of resources. Again, this colder version generates a sentiment through which the situation is approached. This time, the sentiment is infected with the logic of decontextualization and dehumanizing metrics.

The cold questions are compelling, even though they are the wrong questions. These questions distract from the process of perception of the complexity of each person and tuning into the alchemy they produce. As such, they can easily hijack the limited time and morale needed to "find a way."

The dark history: As it turns out, there is a reason this type of approach feels so cold. The creator of the "lifeboat ethic," Garrett Hardin, was a eugenicist, nationalist, and white supremacist. Hardin was an important figure in the history of environmental solutions, especially known for his popularizing of the idea of the "Tragedy of the Commons." The history of Hardin's eugenics and nationalistic thinking has been known all along; it was never a secret. But those ideas and the sentiments under them came to set the tone for how the response to complex crises is imagined and managed. And that is something that requires a sobering wake-up, learning, and new attention to the ways in which that genre of thinking sneaks in. Hardin was not the sole promoter of this thinking; instead, his work is indicative of a culture of thinking that resonated with eugenics. Ironically, adopting a warm data approach to the culture reveals the receptivity to cold ideas (Amend, 2019).

What is being revealed?

As the COVID-19 pandemic so brutally revealed, the institutions of current societies cannot respond to the needs of the people; they cannot meet the complexity. The complexity of the pandemic of 2020 continues to be educational for citizens around the world in how one tiny virus can change the daily patterns of family life completely while it also reframes trust in political, economic, cultural, and medical institutions—at the same time. It also provided a glimpse into the possible collapse of many of these institutions and the gaping need for something in place to provide safety.

It has become clear that the great "they" of societal safety guardianship is not there. Citizens assumed their governments and the institutions their taxes went towards would protect them from COVID-19, but they did not. Reliance on institutions designed with a focus on assigning "value" on the same basis as in the cold lifeboat story has proven inadequate to deal with what matters when confronted with existential crises. Financially and structurally, the ability to help is also simply not there; there is no institutionalized sentiment to care. As crises of economy, loss of biodiversity, climate, culture, mental health, and many more pile up—no one is coming to help the communities of people most in need.

Many conditions have come to form this situation. One is a culture of individualism and separation that has produced a loss of ability to respond together to crises. Independence and a narrative of separation have been ingrained into every phase of life, negating the deeper, more vital forms of interdependence that the future depends upon. While there are cries for collaboration and unity, they are often founded on the grounds of competition, roles, and mechanistic notions of who does what. This is not collaboration, nor is it a living response.

A pre-scripted and rigid set of roles is a harbinger of an era of errors in thinking in which most aspects of daily life were modeled into industrial patterns—from education to health, to the economy, agriculture, even psychotherapy, the grid, and the engineering for ever higher levels of efficiency have permeated our lives. Incentives to compartmentalize one's humanity hold in place artificial distinctions between organizations, institutions, communities, nations, "races," individuals, and on and on it goes.

In contrast, a living response builds relationships, and those relationships go on to build more relationships—making no claims—as the soil does not on the forests or meadows that it nourishes. Warm collaboration is not cloned, not a formula. It is built on what matters in life as well as life-giving and life-supporting values: care and love matter.

At the core is perception. Perception of self, perception of others, perception of the natural world, perception of what it is to be alive. Many of our perceptions are shaped by so-called scientific facts, yet when it matters that we learn from science, we choose to ignore it as an inconvenience. For example, as Mamphela says, "Ancient history and science tell us that there is only one human race, yet we continue to speak of different races."

And now, People NEED People. But to get there, a starkly different approach is needed. And an honest look at current approaches is at hand. There is some not-so-pretty history lurking in the hunt for environmental solutions. As the statues of traders of enslaved people are falling in cities around the world with the Black Lives Matter movement, the need to address the hidden agendas and assumptions and how elitism and control keep manifesting as eco-idealism—is a requirement. Like a detective with a black light, it is time to look at where the blood is to check for fingerprints.

Beginning to embrace new thinking, warmer thinking—in relationships will nourish a new version of solidarity to participate in a world of vibrant, creative, and unimagined change. The Warm Data Lab familiarizes groups of people with the ways in which the contexts of their lives marinate and overlap into each other, and it offers an introduction to the complexity of their own lives so that they may better see the complexity of others. Through this discovery, people begin to see how vital it is to tend to their families, communities, and the land, and they are able to respond to emergencies with warmth.

Commons, Communities, Colonialism: In response to the ecological crises, there has been a push toward addressing what is known as the commons. The benevolence of this work always appeared watertight—help the communities to unify, help them to be sustainable, and help them find continuity. What could be wrong with that? The first question is, "What is a community?" Is it a group that shares a bio-region? Is it a cultural group? Is it a group that shares an experience, such as breast cancer victims or children with autism? Is it a group that shares an activity, such as jogging with dogs or making music? Is it an online community? A shared profession, such as doctors around the world?

So where are the commons? And how can the communities learn in mutuality from within and between the many forms they live within? Community is complex. The multiple contexts of community are not separable; they co-inform one another. A person usually inhabits numerous communities, in layers-on-layers, in relationships with relationships. The eagerness to define community and to define set formulas for responding to the needs of a community is creating a block in perceiving the necessary complexity, perpetuating the elimination of contexts and failing to perceive the uniqueness of the ways in which communities are alive and entangled.

Dangerously, the notion of creating solutions for the commons is often inherently infected with the thinking that the "lifeboat ethic" was forged in. The eugenics is there, implying control, implying scalable solutions, implying that people are numbers and that communities are equations to be solved, measured, and managed. This is Newtonian physics misapplied to living systems, and it may be part of the reason so many regional communities become devitalized in their community gardens, local economies, and so on. The abandonment of community gardens is probably a response to the lack of on-the-ground social solidarity and mutual learning that would choose that garden and generate the relationships that would overlap lots of communities to keep it nourished. The idea of the garden needs to be homegrown, not implemented by planners, no matter how altruistic.

Solutions to scale defy the complexity of the people, the places, the ideas, and the situations. Scale is a trendy concept that must be used with extreme caution. Some projects and products do scale; others do not. The distinction desperately needs attention and articulation. When is the urgency to scale a project a form of colonialism? And when is it not? When is it warm? When is it cold?

For years, we have been troubled by what feels like a cold approach to developing solutions to environmental crises. Something has felt very off. It has taken time to sense and articulate what was amiss, and now that it has become more apparent, there is a need to address it as coherently as possible.

The process of the Warm Data Lab (and its online version, People Need People) is in vibrant relationship with the Letsema Circles of conversation and other cultures of engaging with the interdependency of life. The contrast is between approaches to engaging with complex emergencies, one cold, the other warm. Between them is both the potential to meet the nearly impossible odds of humanity finding a way to co-exist and also a keen look at the great danger of manifesting ecological solutions that serve to continue the thinking of the existing system and perpetuate colonial, cold, eugenic responses to the present and coming emergencies around the globe.

Hope lies in the very fact that, as living beings, we are wired for relationship. It is only possible to express our humanity in relationship to other human beings. We exude warm data in our eyes, our smiles, our verbal and non-verbal conversations, and the importance we accord to relationships with others at the personal, family, and community levels and in local, national, and global contexts. We also can tap into ancient wisdom to add to our store of warm data and the capabilities unleashed when I see you in me and me in you. Our well-being and that of our planet is possible only if we permit ourselves to perceive and embrace the rich expressions of who we are as living human communities and to find a way in relationship. This will require warmth and rigorous attention to relational integrity above the anxiety to control.

* Dr. Mamphela Ramphele has been a student activist, medical doctor, community organizer, researcher, university executive, global public servant and is now an active citizen and a trustee of the Nelson Mandela Foundation. She authored several books and publications on socio-economic issues in South Africa. She has received numerous awards for her work for disadvantaged people in South Africa and elsewhere.

TONE

When culture is disassembling, it is also
reassembling.
The tone, the aesthetic, the language, and
mutual care is what will FORM and in-FORM the
connective tissue
of the new linkings.
The tone matters.
Kindness is a tone.

TRAVELING ON A PAVED ROAD

At speed.
At scale.
The illusions are wearing thin,
The distance covered reveals foolishness,
Wasting & wanting & wounding.
Taking ... taking ... taking.
At speed.
At scale.
No way to put it back,
to un-waste, un-want, un-hurt.
Just come close,
Slowly.
Unevenly.

SOMEHOW

Somehow we have to carry trauma and stay imaginative enough to rebuild a society that cares.

Somehow we have to be angry enough to fight and tender enough to stay sensitive to the nuances of mutual learning.

Somehow from the fragmentation of human constructs, we have to find a way to perceive and participate in the unity of life.

Somehow we have to.

STRETCHING EDGES

It takes all aspects of you to meet the world.
It takes requisite variety. Sensorial versions overlapping.
A million tiny hairs, and a repertoire of sound shapes,
wrapped in memory and
hiding deep below the conscious verbal world
Attach and taste each encounter, skipping across time.

I can't remember what we talked about,

Fig. 14. (pp. 78-85). Rachel Hentsch (design, composition), with Nora Bateson (photography, poetry text). (2023). *Stretching Edges*. [digital art].

Where art lives.
In that soil the senses
are refound, reframed,
remade, resung,
redrawn, rewritten,
rotting, and reconnecting
to new organisms,
new ideas.

but I remember the texture of your words.

When you meet another person, how do you make sense of them? Do you measure them and look to their statistics?
When you hear a piece of music or enter a forest, how do you make sense of it?

What languages are you fluent in?
What can you read in a person or a song or a forest?

How can we describe what we cannot perceive, or perceive what we cannot describe? Are there new textures of being yet unfound?

I am not somewhere outside of ART.
There is no way to breathe
or eat or love
or move your body
or take part in a culture
or drive a car
or pet a dog
without doing it:
art.

Where the meaning-making is alive
not concrete, where the alchemy of information
is always stirring in history, language,
pain, desire, humor,
beauty,
rage.

A baby cries,
A cat purrs,
A tear streams
down
a friend's
cheek.

Where do we
understand these
things? They are not
pluckable stats.

They are not direct.

What if the caterpillar were looking for a way

I can't remember what we talked about, but I remember the texture of your words.

to design sustainable everlasting caterpillar-ness because it could not,

because of course it can't,

imagine itself as a butterfly?

SELF PORTRAIT

If I were a mathematical formula, I would be a chalkboard full of symbols and arrows. No ... I would be many such chalkboards. Or maybe just a crayon writing 1+1=
With no answer.
Wondering.

Perhaps as a song, I would be a combining of something that resonates in thunderous bass tones with a lightness on top and a thumping beat ... There would be a chorus of punk rock as well as a clear ping-ping of water on a ceramic bowl.
Dancing.

Or, as a meal, I could be, on some days, a handmade feast of carefully overlapping tastes, textures, and temperatures. Salty and sweet, spicy and bitter. Singe your tongue in one bite, cool drink of bubbly in the next. Other days, I am reheated leftovers soaked in yesterday's flavors.
Reminding.

As a weaving, I might have threads of moss, and my beloved's hair, the purple plastic tassel from my bike when I was nine, the silk of longing, and itchy wool because life is many things but not comfortable. I would be an elfish cloak that allows for air and magically keeps out the cold. Wrapping you in love and ideas and giving you gumption to explore.
Tending.

I could be a painting with thick chunks of color and contrasts, broad shapes that say BANG! Or the tiniest, daintiest filigree of small brush strokes, a blade of grass arching in a breath of wind. An ink on paper where the ink soaked through to where we meet.
Reaching.

I could be a poem, like water that reflects the me that you see and changes each time you read it. Crafted in language so open that I remain able to move and change within the words. In that case, I might not make sense.
Learning.

I could be a forest, a meadow, or a tide pool. A desert or a tundra. I could be a puddle of muck that is just forming into moist possibility. I could be a mushroom holding the communication between trees.
Living.

In each of these, I am described in a set of messages that explain and confirm one another.
Messages that sing to each other.

And when we meet, you will see another set of images, a woman in shoes and accessorized in the thing-ish world we live in. I will dress for the occasion of our meeting in something bought in a store that covers me in a culture of signals so you will know where to put me in your library of codes.
And you will do the same.

Hundreds of years of messages are flying between us ...
Respectable?
Lovable?
Worthy?
Status?

But which codes have we chosen to message each other within?
Which signals are obscuring the others?

Can you still feel the texture of my math?
Can you breathe the ink of my wool?
Can you be a blade of grass with me?

Implicitly, in the waft of us, in the style of us,
In the way we are ...
Is the possibility of our communication.

Changing the code changes more than the message.

These are portraits of possibility.

Combining

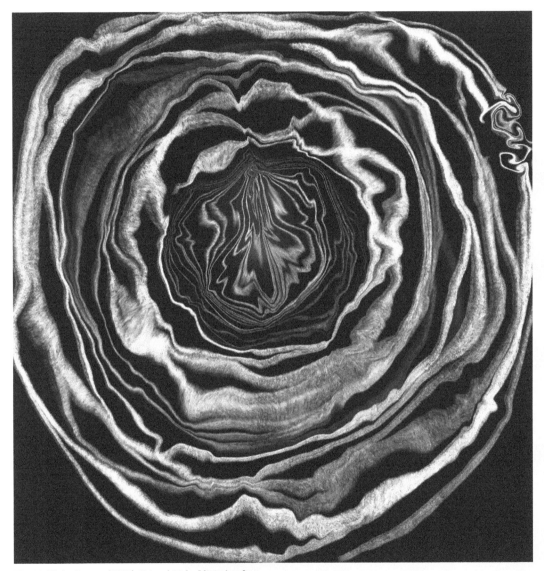

Fig. 15. Vivien Leung. (2023). *Fainted Light*. [digital art].

Fig. 16. Vivien Leung. (2023). *What We Perceive*. [digital art].

COMBINING

Every hole is a story,

every patch has a story.

Every wound is a story,

every healing is a story.

I LOVE YOU

I will meet you in the midst. Unready and lost,
but with the nectar of possibility in my mouth.

Each day there are horrors unspeakable, baked
long in histories, unchangeable,
and life is still outrageous.

Each day, despair is undeniable -
destruction steeped in justifications deeply
motivated for why this is ... this violence
should not be.
and life is still untamed.

Each day there are plenty of reasons to give in
and give up - the effort to keep going runs dry,
and life is still unstoppable.

Each day I wake up
and life is there, singing through the impossible.

This is goodness unfettered, meeting the enormity
of this mess with nothing other than
skin-bare
banality
of the sacred love life has in its pulse,
bringing one day like a storm
into the breath of the next.

Combining

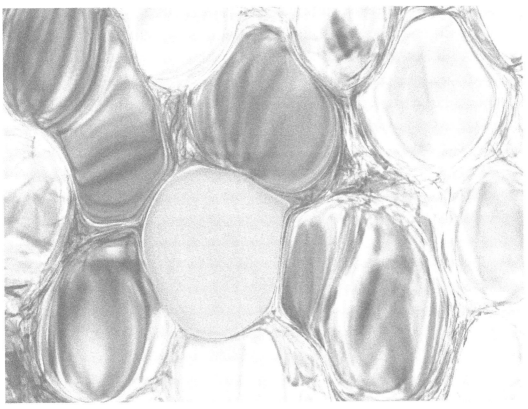

Fig. 17. Rachel Hentsch. (2023). *Fantastic Six Blobs*. [digital art].

WITHOUT GOING BLANK

No matter the outcome tomorrow, the work
remains the same,

To touch and hold the coming years in warmth.

To carry the brokenness in hands full of
flavors,
hearts in full color.

The tending is the absolute.

Ardor is the fire
– of meeting the day without going blank.

Combining

Now

Now, when the reveal of the damage done makes for an enormous Shoulda-Coulda-Woulda, maybe it's possible to have:

Been kinder
Woken up earlier
Been more careful
Demanded justice
Been more outspoken
Used more imagination to tackle things that were thought to be hopeless, idealistic, impossible . . .

I am not one of those people who think it's best just to let the past go. I grew up around positive thinkers who repeatedly did some of the most absurd and myopically hurtful things you can imagine. In the name of self-care and strange, more-bliss-than-thou righteousness, an awful lot of selfishness has taken place and been justified as "the best we could at the time"—I have a rough time with that line.

I prefer to dig deep and wide to search for perspectives I may have missed. I am compelled from within to find the scope and the shape of the illusions I have fallen prey to. I revisit my mistakes. Not to harp on them or wallow in inane puritanical guilt, but to find the oversights and erasures. Or at least try.

I am not a forgive-and-forget person, especially when my mistakes have hurt people. I want to rigorously study and learn what happened in me, in my situation, between us. I want not to repeat the issue. I do not want to be forgiven; I want to expand my perception so that I might become sensitive to whatever I did not see that allowed me to injure another person. It is easier to see what went wrong later on. As the contexts come into view over time, it is possible to make sense of the mistake with what appears to be new information. But it's not new. That context was there all the time; it was just invisible. And the learning does not stop. As time passes, insight expands, and even more context comes into view.

Sensitivity grows—like learning to tell the difference between running on the grass at the park and not running in your grandmother's newly sprouting flower garden. It is possible to enhance the receptive systems to take in dangers that are one step removed from my own immediate experience. They are not my flowers; maybe I don't even care for flowers, but I care for my grandma. I may overlook the working conditions of the people who sew or dye my clothing. I may not notice the die-off of the soil bacteria due to the chemicals in my kitchen. I may not be concerned about the gathering of other people's personal information through their tech devices . . . not my problem. Until one day, I have another sensitivity, one that recognizes, in hindsight, that their freedom, health, and vitality were, in fact, inextricably tied to my own.

It is possible to perceive more context. A fisher who fishes in a particular river knows how to read the ripples in the water's current, knows the seasons, knows without knowing they know when and where to drop the line and with what sort of lure or bait. The conditions change from hour to hour, week to month, but their sensitivity and ability to read the context is an ever-expanding process of learning. Raising children, I have learned that there are times when I "know" to be more stern than other times; the same infraction can produce very different responses from me based on the day, the circumstance, and the child's emotional resilience in that moment. We pay attention, we adjust, and we learn.

For me, showing up now is about not shying away from the horrors of the past. But, also not basking in the pain of being party to those horrors—which is equally self-serving. I have benefited from the exploitation of others and the natural world. Sick, but true. There is nothing to gain by justifying or denying that fact. My comfort has been made possible by the suffering of others. Generations of pain, trauma, and confusion have been produced by the systems that give me my food, clothing, transportation, banking, and everything else in my world. There is no point in pretending that is not true. Even though my family has advocated for the interdependency of life across many fields for at least three generations, I am still dressed in blood. The illusions that have made this life are ghosting around me; they always will be.

So, now what? All I know to do is to listen, to watch carefully, to keep trying to make sense from different angles. One thing I am starting to get the hang of is igniting suspicion toward those impulsive solutions—which appear at first to make the most logical and rational sense—recognizing that they are probably informed by the same systems that created the problems to begin with. I am in rapid wide-angle receptivity mode—all senses on. I am watching the scaffolding of the last century of exploitation pixelate and melt.

I am sad, mad, lost, sorry, and grateful. Ready to scream louder, touch softer, think harder, run faster, watch more carefully. Care, fully.

Do not forgive me; do not forget what has happened. It is me who must make the move. I will. I will stay alert, stay learning, and stay honest to not-knowing. I am here now to tend, to offer, to generate, to witness.

Hindsight is sliding around on the slopes of time—not just looking back, but forward as well.

Hindsight is looking into the eyes of future generations without excuses.

WHAT I LEARNED

I have been seeing one small piece of the landscape,
A branch of a forest.
Believing it.
Then —
The forest revealed,
And I feel I have been so inept.

It is a reminder of the cost of thinking my knowing is anything.

I am learning that learning is the healing —
Not backtracking, not re-hashing, not splitting words, and attempting to save face, not apologizing.

The only currency here is movement.

Learning is moving to a scope from which there is no return to past limited vistas —
No going back to not seeing,
No hiding in old ways of non-sensing.

It is a one-way valve.
That changes everything.
It changes me.

I am learning to see the outlines of my own misplaced knowings.
I have learned to watch for signs of cocky with the stance of clarity.

The self that thinks —
I already knew —
Is a version that cannot offer anything new ...
Now when that version of the self shows up,
I will recognize her.

I learned that learning is a relief –
like oxygen rushing to limbs that were in bondage.

I am learning to question the images and the stories I was told, to smell the blood in them.

And learning to spot the frames, bars, bondage so camouflaged into the culture.

I am learning to not hope to solve the past harms.

The pain that sits there just sits there.
Caught between now and four hundred years of torn lives –
Nostalgia has no place here,
It is a blade through the scripts that obscured and justified.

I now know better than to think I could see this history.
Any history.
I do not see. I cannot know.
My effort will never be complete.
It will never count. It will never be enough.
I will hold in open rigor
the ongoing learning.
But I am able to sense a little bit more than before.

Learning to balance in the instability of a wall-less, floor-less world.
Seeing one small piece of the communication,
A single word in a whole language.
Believing it.
Then –
The conversation revealed,
And it was time to be silent.

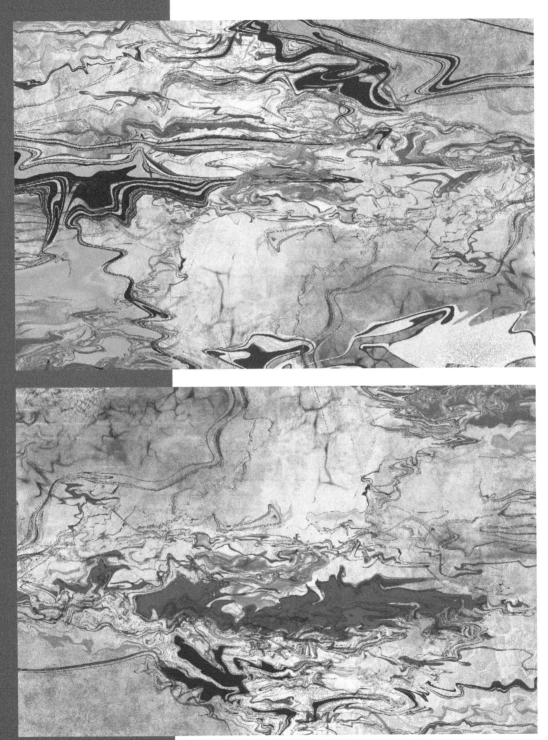

Fig. 18. Leslie Thulin. (2022). *Always Moving*. [digital art].

It's Fantastic

It's fantastic.

Just to witness such a theater of tangled stories is indeed a show to beat all shows.

The way each flavor of the stories of this moment has produced its own puzzle, complete with transcontextual overlapping versions is remarkable, even terrible, but beautiful in its intricacy.

It is fantastic. It is everywhere. It is a whirling maze of reiterations. A fugue of seductive cages for epistemology to stir itself into itself.

Through one door, it's religions; a tangle of the sacred, the potential for devotion wrapped into a spiral of exclusions and controls . . . and God is god because of god, and god is absolute.

Through another door, it's money, and money is real because what is real is defined in terms of money. The buck stops and starts there.

Then there is the academy, and what is real is studied because what is studied is real. Research proves the research.

The schools are labeling, measuring, and funneling the next generation into compartments that desensitize the ability to receive information about life.

The law is based on ownership, which is based on law.

The machines keep telling us we need more machines.

The health systems are making people sick. "Pharma" is a lifeline that is laced with lies.

Self-help is producing an unhealthy idea of the self, and help.

Spirituality . . . is full of wordy scripts where only breath, art, and music can offer communion. Words are inadequate.

Combining

So, they point to imagination for a break from the boxes, only to find that the imagination is sourcing from the epistemology that produces boxes. Remodeling the models.

I might shrug my shoulders, sigh, and think... well, it is what it is.

As Korzybski says... it IS not what it is. And that is the fantastic part.

The perception of any one of these tautologies is forged and patterned into a writhing cluster of habits and language, spinning round and round. The thing seen is not the thing. The thing felt is not the thing.

Caught in one thousand traps at the same time, at some point, one must admire their magnificent mire. All of the stories are in cahoots, even in their contradictions—proving the others in a world of proofs.

The education system proves the economic system, proves the political system, proves the culture proves the health system, proves the religion, proves the parenting of the next generation into the same traps. And around we go, never touching ground.

It's fantastic. All attempts to wiggle it, upturn it, reverse it, or destroy it are thwarted by the momentum. You can't change it because it is what it is. Even though it isn't. It is.

What a tragic comedy, a romantic adventure, a fun house of optical illusions. Not fun.
It's nothing short of incredible.

Now, hold, pull, release. Let the anxiety of each failure to make change pile on top of each failure to perceive another way until you can climb on top of them and peek out.

Here it is, the slippery stuck-ness. In its redundancy is its continuity.

In the continuity is the ever-shifting, in the ball of interlocking snakes of story is the way out. Fix one story at a time and dig your way deeper into the tangle. You have to dance with them all at once.

You have to dance them out of their gripping of each other. I have to dance you out of your embracing me into this stranglehold. You have to dance me out of my idea of who I am when I am with you. Let me be unpredictable.

The stories tell each other.

The institutions reflect each other.

The traps are not simply locked doors, but rather, they fasten one another into the sphere of reconfirmation. They whisper in hissy voices... thisssss issss realllll... You are real because I am real.

War is human because history has war. Offshore accounts are possible because there is no law to make them impossible because it is possible that offshore accounts rely on impossibility. The market is driven by exploitative labor because the labor needs the market that exploits.

Shattered mirrors reflecting strange angles of the same crime scene.

It takes humor, art, reverence, and irreverence . . . it takes rigor, practice, stretching the brain. It takes synesthesia, sensing in new ways across senses. It takes warmth, weirdness, and wonder.

I feel I am screaming in space, writing in invisible ink. I can tell you ahead of time that I am going to seem incoherent, un-informed, irrational. I am an aberration, a nonsense. . . . But that is what change looks like.

I mean, seriously, this is why play is vital. The change needed looks unfamiliar, looks nothing like the existing perceptions, looks nothing like what might be called grace or profanity from this vista. You cannot trust yourself to recognize it when it kisses you. We cannot rely on this current vantage point to reveal the change—we cannot see it or name it from here. It is not in existing language. There are no subtitles.

It is crickets, wind, light waves, and the call of next year's baby goats. We do not hear it. And when we do, it sounds too strange.

So where is the change? It is everywhere except where we try to make it. The trying and the making of change are contaminated with the familiar scripts and blueprints. Watch the swirl—try not to be distracted by what is swirling.

The tautology is a tautology.

It's fantastic.

Fig. 19. Leslie Thulin. (2023). *Together*. [digital art].

Simultaneously Implicating

I have a meadow. It's not a meadow out there in the landscape. I don't actually have it. I just keep it with me. It pushes me into un-stagnant, unfrozen, un-oriented sensitivity. It loosens the grasp of the ideas pulled taut against the movement of life.

The meadow is in me. I return to it to remind myself not to get trapped in all the gripping notions of what life is and isn't in this era. I get sucked into the logic of things like economic reports, voicemail choices, and the laws of immigration. They start to seem real. The meadow is where I am reminded of the edges of my perception and challenge the trickery of my well-worn inclinations. I practice thinking like an ecology there. I am studying the ways of it, wondering:

How do things stay the same while they change?

> If all the organisms in an ecology were absolutely constrained to their roles, there would be no communication, no movement, no evolution, no life. But, if all the organisms in an ecology were in wild flux, they would be unable to remain an ecology.

It is dangerously easy to see ecology as a machine. If a system is systematized as a functional collection of roles and purposes—you are likely looking at more machine than organism. The organism will break the rules of roles. While order is needed to ensure that the ecology can stay alive given the thresholds of the needs of each organism, learning is necessary to ensure that it is possible to stay alive through variations of environments, crises, and events.

The ecology tilts toward consistency, shutting out new information to stay alive.
The ecology also generates new information, averting the obsolescence of consistency to stay alive.

What stays the same, and what changes? What is conserved matters. Each organism is shaped by the others in its environment and is shaping the others simultaneously. They constantly implicate one another in an ongoing mutual learning (symmathesy) oriented toward continuing vitality. The savanna, the jungle, the sea, the forest, and the tundra change together, keeping the larger ecology in play. The premises of the ecosystem defer to life. But the premises of present-day human socio-economic systems defer to themselves. There is a disconnect.

Combining

I am certain that until the perception of tensions in these ecological phenomena is practiced in as many contexts as possible, any attempts at "changing" will serve to perpetuate and worsen what is sometimes called a polycrisis.

A polycrisis is only makeable in an ecological way. It is an ecological consequence of ecological consequences. It can, therefore, only be unmade by being met (not matched) in new ecologies of connection. The economic crises contribute to political and health crises, exacerbated by technological crises—while ecological crises manifest in health crises, which contribute to economic and political crises—all of which produce pressures on education systems and family crises. What is produced is an ecology of crises, all interdependent and inseparable.

The polycrisis is the product of too many looping traps inherent in the "premises" of the interactions in the system. In the case of "modern" human social systems, we see a circling of basic needs like housing and food linked to economic need, which is contingent upon ongoing growth of technological advancement of everything from agriculture to pharmaceutical research, which political bodies must perpetuate and protect through military-industrial development and support with financial systems that are not possible without workers, materials, and transportation. . . . All of these require what the oceans, rivers, forests, geology, soil, and animal life offer. The premise is one of staying viable in the existing systems, which are not viable in the ecological systems.

When things are unchangeable because changing them is not possible, you are likely looking at a tautology. And it's not possible because they are unchangeable; it's fantastic. A term for this loop is "tautology"—it refers to a closed circuit. In a tautology, the information is the basis of the information. Sometimes, it is thought of as saying the same thing in various ways. "Tautos" is Greek for "same," and "logos" can mean word or speech. It can also mean "same explanation," such as the way in which Euclidean geometry is proven by Euclidean geometry or research proves research.

The critical thing to note is that it is not possible to add new information to a tautology. It is true to itself—itself. It is an informational eddy, or whirlpool, that loops in on itself. Without outside input, it spins endlessly in its own circles. It is what it is. Interestingly, the person who brought tautology into logic and math was Charles Sanders Peirce, who also introduced abductive process. Peirce said, "Abduction is the process of forming an explanatory hypothesis. It is the only logical operation that introduces any new idea" (Peirce, Houser, & Kloesel, 1998, p. 216).

Abductive process is the way in which the information derived in one context becomes the basis of information in another context, which may not appear to be in correlation. Peirce introduced the notion of hypothesis in abductive process. (i.e., the experience in one context gives information about another.) For example, after going through the school systems, students can hypothesize what is wanted from them on forms and surveys in other contexts like health or social service institutions.

My father was also interested in abductive process and saw it slightly differently. He was interested in the way that what is learned in one context actually becomes a description of another. Such as how students have learned to make sense of the school system becomes a description of how they will learn to make sense of other systems like health or social services. The difference is slight but significant. In either case, the importance of the way the information is placed elsewhere creates something like a metaphor. Metaphors are not closed circuits. Metaphors are loose; they have gaps; they invite overlaps and communication variables.

TAUTOLOGY is tautology. It has no gaps. No new information can get in.

Abductive process makes gaps. In the "mismatch" and overlaps, new information gets in. And I find it both very useful and amusing, and sometimes frustrating to note how many tautologies are spun through language and culture that perpetuate all sorts of fallacies. The most common one is that reality, as current systems have described it, is "real" simply because it is described as such. (i.e., it's real because it's real; therefore, it's real.) The next logical step of that tautology is: since what is real is really real, it is natural and cannot be changed.

These circular loops are fascinating but deadly: Perception, deception, bias, buy-in, illusion, profusion.

You probably know this word, and knowing the word makes it possible to know the word. Tautology is a statement that is true by necessity or by virtue of its logical form.

But, it is NOT what it is. It is so much more: a whirl of conditions and relations.

While words often help us see, they also deceive us and perpetuate illusions that keep the tautologies taut. Words tie us into knots and guide our thoughts. Words also crack open the "thought knots." I am inviting this word into the situation room. Without understanding how the tautologies are fastening themselves, each attempt at loosening twists makes them worse. Tautologies close tight and are caught in their own logical locks. In contrast, abductive process offers the possibility of asymmetry and mistaken responses, allowing new information to get in. Ecologies are both prone to stuck-ness and always changing.

> My opinion is that the Creatura (living systems), the world of mental process is both tautological and ecological. I mean that it is a slowly self-healing tautology. Left to itself any large piece of Creatura (living systems) will tend to settle toward tautology, that is, toward *internal consistency* of ideas and processes. But every now and then, the consistency gets torn; the tautology breaks up like the surface of a pond when a stone is thrown into it. Then the tautology slowly but immediately starts to heal. And the healing may be ruthless. Whole species may be exterminated in the process. (G. Bateson, 2002, p. 194)

I slip—I get caught in the net. The way my meadow's organisms are living among one another allows the meadow to keep being a meadow. The organisms are not falling together into an accidental envisioning of the workings of a watch—only squishy in its mechanism. Life is not like that. And life is like that.

Combining

In my meadow, I sometimes pretend to be an ant—with legs at my sides—crawling through the plants looming over me. Or I picture myself as a grass stalk, feeling the stretch and weight of being long and bending in the breezes. I go under the meadow floor, into the grit and cool of the earth. I attempt to be tiny—as small as bacteria—then as heavy as the tree whose shade the moss loves. As I play with the possibilities of "umwelt," I am contrasting, tracing over, speculating, and changing vocabularies of experience from ant to grass to moth. There are overlays that confirm and undercurrents that contradict. There are layers of gaps forming. I am un-cutting my own language, reaching into the metaphor.

The meadow is in my head, totally concocted within the restraints of being me. My fantasies and the limits of my experience contaminate it. I have never been an aphid or a moss.

My visits are more than isolated fantasy—I go there guided by my senses, memories, and irrationalities. I return to this meadow a little differently, and the meadow is also a little different. Time, ecology, recursiveness, and evolution are all in motion . . . trying on the meadow in variations. At times, I am my unborn great-grandchildren lying back on the hillside watching clouds. Or my great-grandmother when she was a girl, lying back on the hillside, watching clouds. Through the meadow's continuation, I am able to host time as a guest in many shapes. I am borrowing senses I do not remember but that my membership in life has given.

In 1888, my grandfather said his "brain boiled with evolution" [see page 172]. All these decades later, I find am with him. The impossible possible of my meadow is a constant reminder to pay attention in many textures, tones, and sizes—allowing my senses to describe new senses.

My meadow is relentless in its aptitude for multi-tasking. It sows seeds in bird poop while decomposing leaves, housing rodents, giving the spiders blades of grass to make their webs in—all the organisms are unstill together. Simultaneously, they are implicating one another into viability, vigor, convivial conversation.

The color of the aphids is simultaneously implicated with the stems of the flowers; the permeable skin of the earthworm is reflecting, describing, responding to the rainfall; the roots are holding the soil; the minerals of centuries of composing and decomposing. . . . The organisms are shaping each other even as they are changing shapes. The organisms are explaining one another into their ecological continuing.

> Last, there is a special form of "knowing" which is usually regarded as adaptation rather than information. A shark is beautifully shaped for locomotion in water, but the genome of the shark surely does not contain direct information about hydrodynamics. Rather, the genome must be supposed to contain information or instructions which are the complement of hydrodynamics. Not hydrodynamics, but what hydrodynamics requires, has been built up in the shark's genome. Similarly, a migratory bird perhaps does not know the way to its destination in any of the senses outlined above, but the bird may contain the complementary instructions necessary to cause it to fly right. (G. Bateson, 2000, p. 143)

"Simultaneous implicating" is a phrase that brings me into the movement of ecology. The concept of ecology cannot be a parking spot or a static definition of interdependency. No, that is not enough. The organisms in my meadow and the shark in the ocean are both shaping and being shaped by the other organisms they live with, who are also shaping and being shaped. The bees are implicating the flowers, the birds, the grasses, the butterflies, and the bacteria underground. The grouper fish and the seagrasses, the microalgae and the reefs are mutually simultaneously implicating one another toward the health of the oceans, or the lack of it. Stuck-ness is also always a simultaneous implicating process, sometimes into mutually reinforcing double binds.

Practicing this perception of simultaneous implicating in each moment is perhaps a practice that is never complete either. But it is an expression of my affection for life. It is also my rebellion against those habits that would overlook the magnificence of this inter-wrapping, entwining, open-ended movement of being a living organism in a world of living organisms simultaneously winding themselves into each other.

The organisms overlap. They all decompose. They have rhythms—they move. Each organism has processes that make it possible to get through events of change and stress, which overlap in their ecological membership. Their redundancy provides compensatory flexibility. It is vital to the movement through time and seasons, events of drought, fire, cold, overpopulation, and extinctions. They finish each other's ecological sentences. The meadow is only possible because they are all implicating each other simultaneously.

Industrial redundancy is another thing. Living redundancy has as its core the need for all the overlaps to be always slightly, sometimes significantly different, and thus gap-able. The particulars of the way the bees and the other pollinators, like butterflies and moths, differ in doing what they do are both unique and repetitive. Repetition in life is there to make it possible for mutual learning to happen and for organisms to compensate for each other in times of stress.

The loop is not the same as the redundancy. Redundancy with variation is how life does it. Ready for stresses and events, ready to learn and evolve, there is patterning, but the fluidity is crucial. A machine or a factory can depend upon repetition, but living things rely upon enmeshed compensatory possibility. To be simultaneously implicating is not a static process; it requires movement in detail.

The ways in which ecologies remain in ongoing consistency is through the possibility of new information getting into the relationships. This is a risk. It could and does lead to extinctions and obsolescence—the greater risk is locking in and looking to repetition instead of redundancy.

Why does this matter? It matters because the thinking about addressing the global crises is currently caught in offices and committees trying to produce industrial, decontextualized, abstracted solutions, formulas, and models for socio-ecological-political situations. This is a tautological response within the system of our everyday tautologies—a proven solution from the history of solutions that create ecological problems.

Combining

The redundancy, by contrast, is living at the detail level of everyday life. Each family and community needs to be able to try to find new ways of making it possible to live. Their ways of being simultaneously implicated with each other and their geographies are particular. They are unique, and their versions of responses should be unique. Each group will have different approaches and feature various contributions to their geographies. There will be varying satisfaction with their attempts; they will try again.... There will be confusion and mismatches. New information will arrive; this is ecological, and therefore, it does not fit into any political or economic structure that suits present governments, financial institutions, or even social services.

The shift is so radical. That is where my meadow is relentless. It does not let me get away with it when my inner industrialist attempts to jump the ecological step and propose another hack job. Fixing one problem at a time is an illusory jump; nature never does that. Even fixing ten or a hundred problems at a time is delusional because it happens in nature at nth-order. Attempting to produce more oxygen by planting trees is a violent misunderstanding of the long ecology of forests—this is an example of a hack job. However, closing a sea to industrial fishing and shipping and making it a wild reserve allows multiple forms of redundancy, simultaneously implicating and re-implicating, to take place, bringing vitality.

The meadow is a constant reminder of the nth-orders of change. For example, in tree-planting, there is a doing to the natural nth-order processes. In contrast, in the closure of shipping and fishing, the doing is to the industrial processes, removing their interruption of the ocean while allowing the ocean to get on with its nth-order simultaneously implicating. It is more like a removal of a hack job.

In algebra class, we learned about variables, but those variables were defined; they had values that did not move. The movement among organisms is different. The values are not defined—they are warping the models in their wildness. Their math is impossible. They are simultaneously moving in different timelines; they are asymmetrical while also reaching toward symmetry. The abductive process is non-stop enmeshing and re-morphing. The meadow is not a model; it is not what it is.

LIFE IS ART

Seasons change everything while breaking nothing.
The mud in the garden, the tiny buds,
the unopened blossoms –
the thought of everything and nothing,
changing and not breaking is overwhelming –
it is extreme in its complexity –
and it is absolutely simple, even banal ...
and beautiful.

Combining

SYMMATHESY

Each one of us is a crooked tree,
Reaching for water and light,
Bending ourselves around obstacles,
Scary thoughts, hurtful moments,
darkness and thirst,
Finding a way to breathe in the sun and hold
the soil,
Our branches are kinked and twisted,
Because that is what it took to be here,
The ways of learning to be in our worlds,
Have shaped responses,
Our many experiences are speaking through every
gesture.
Our loves, broken paths, a tenderness,
a criticism,
Learning always,
Yearning always,
In crooked beauty ...
To be a home for those who may find comfort,
In the asymmetry of our belonging,
A nest cradling new life,
Tucked into an old log teeming with creatures,
learning to be in each other's reshaping.

REUNION*

A thickening of the unsaid integrity –
Starting in small fringes that link and recircuit,
finding unfound mixtures.

Re-soaking the past.
Marinating memories,
Until their softness is sticky vitality.

Like the richness of soil,
The ensembling is teeming with nuances
moiré-ing into other nuances,
Following entirely un-drawable paths.

The unusual textures, the surprises,
in the wordless sea of how we are.

The resonances and rhythms have their own
current in the rich probiotics of fresh tones.
Made together, without goals.

This is not collaboration, this is composting.
This kind of new life is not a restructure.

It is a reunion.
It is not a plan.
It is a nourishing.

*Previously published in *Unpsychology Magazine*

Fig. 20. (pp. 110-119). Rachel Hentsch. (Collection: curation, design, layout, composition). (2023). *Crossing Borders: Families in Motion*. [digital art].

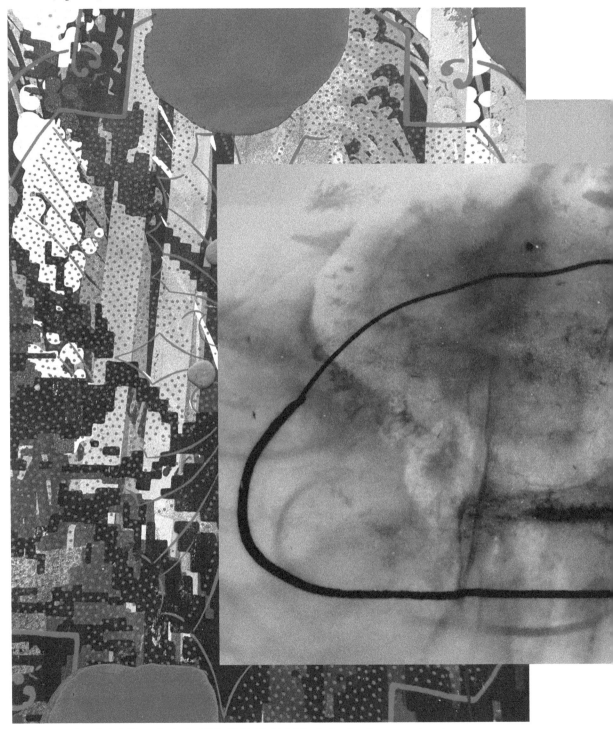

Fig. 20a. Mats Qvarfordt & Trevor Brubeck. (2020–2023). *Untitled*. [Handprinted wallpaper test sheets].

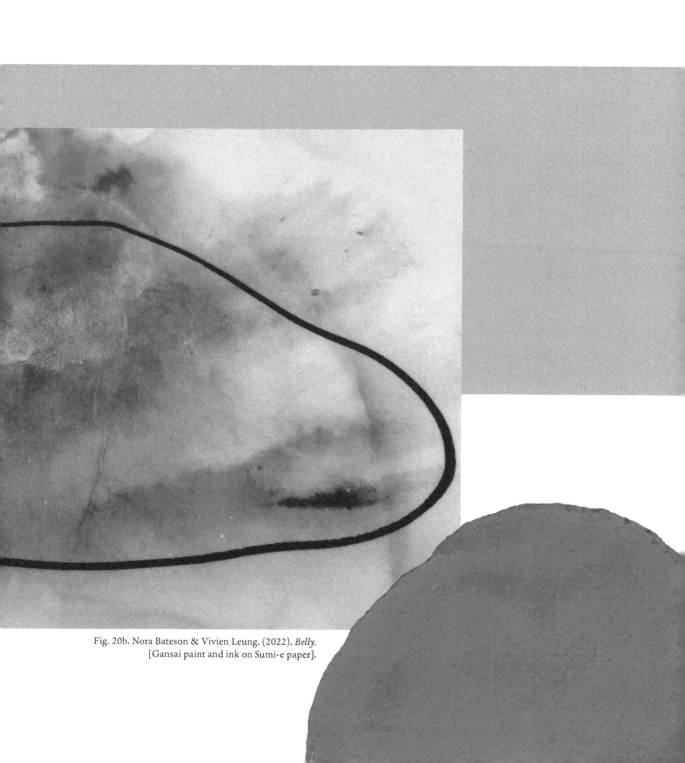

Fig. 20b. Nora Bateson & Vivien Leung. (2022). *Belly*. [Gansai paint and ink on Sumi-e paper].

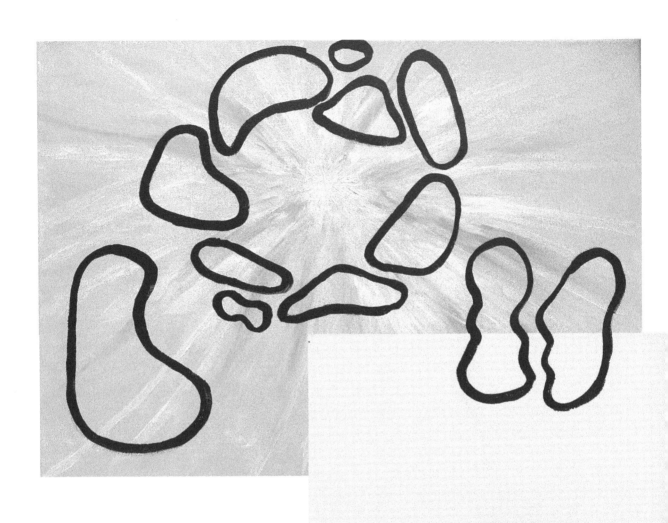

Fig. 20c. Nora Bateson. (2023). *Blobs Collection: Golden*. [acrylic on canvas].

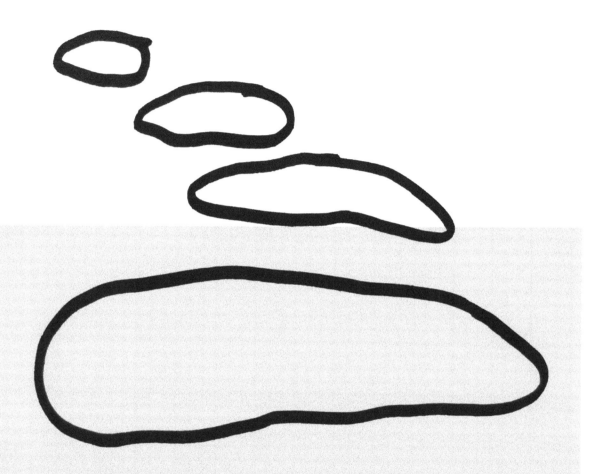

Fig. 20d. Nora Bateson. (2022). *Blob Pathway*. [ink on paper].

Fig. 20e. Nora Bateson. (2023). *Family*. [oil on canvas].

Fig. 20f. Mats Qvarfordt & Trevor Brubeck. (2020–2023). *Untitled*. [Handprinted wallpaper test sheets].

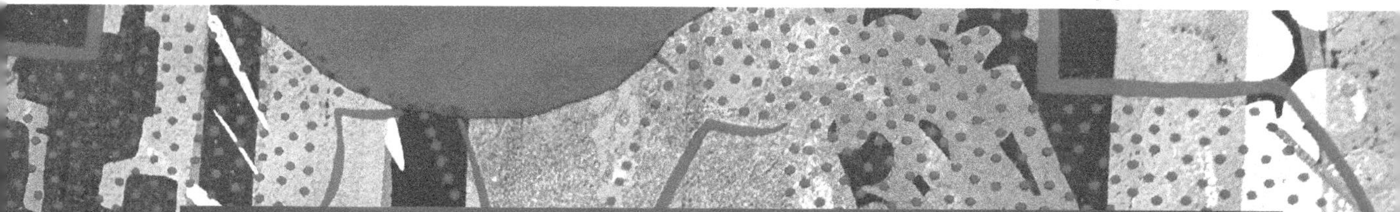

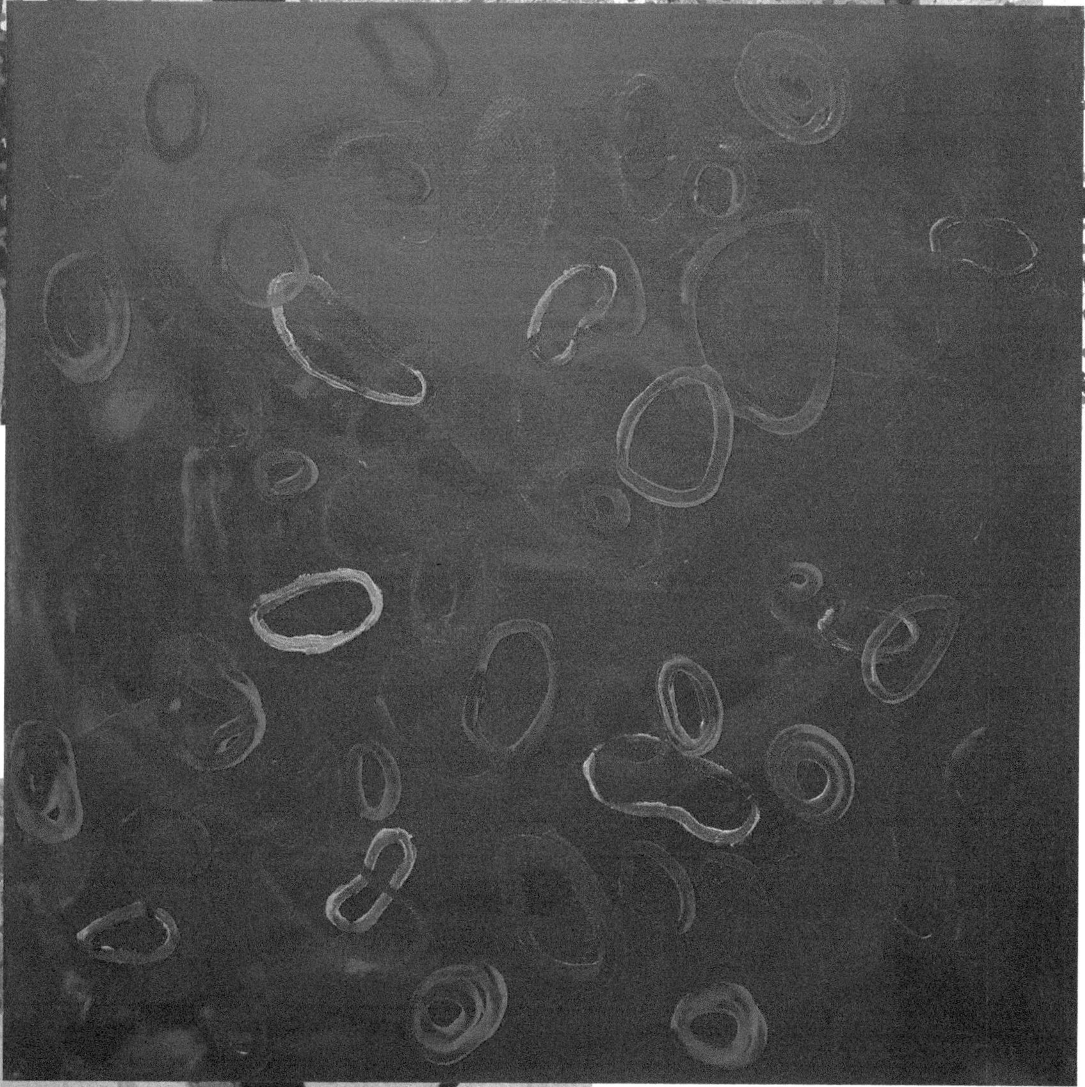

Fig. 20g. Nora Bateson. (2023). *Biology Blob*. [oil on canvas].

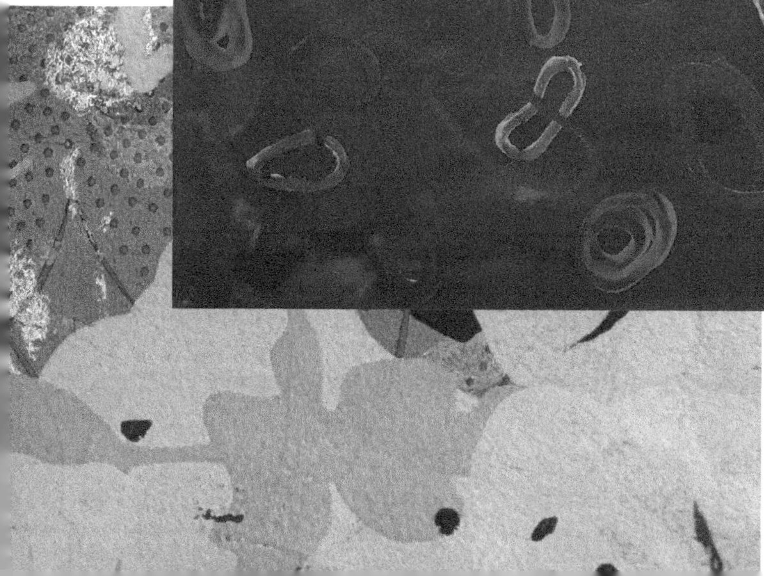

Fig. 20h Mats Qvarfordt & Trevor Brubeck. (2020–2023). *Untitled*. [Handprinted wallpaper test sheets].

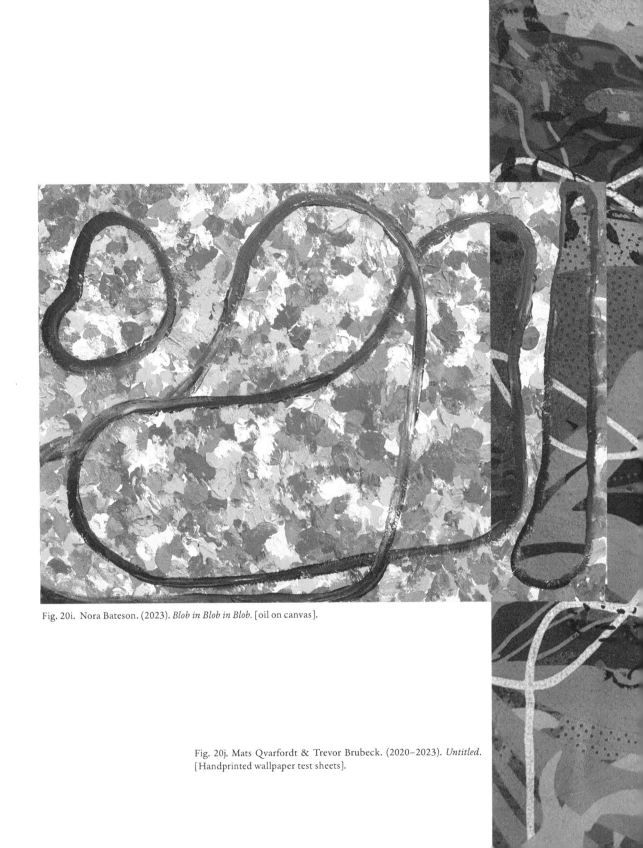

Fig. 20i. Nora Bateson. (2023). *Blob in Blob in Blob*. [oil on canvas].

Fig. 20j. Mats Qvarfordt & Trevor Brubeck. (2020–2023). *Untitled*. [Handprinted wallpaper test sheets].

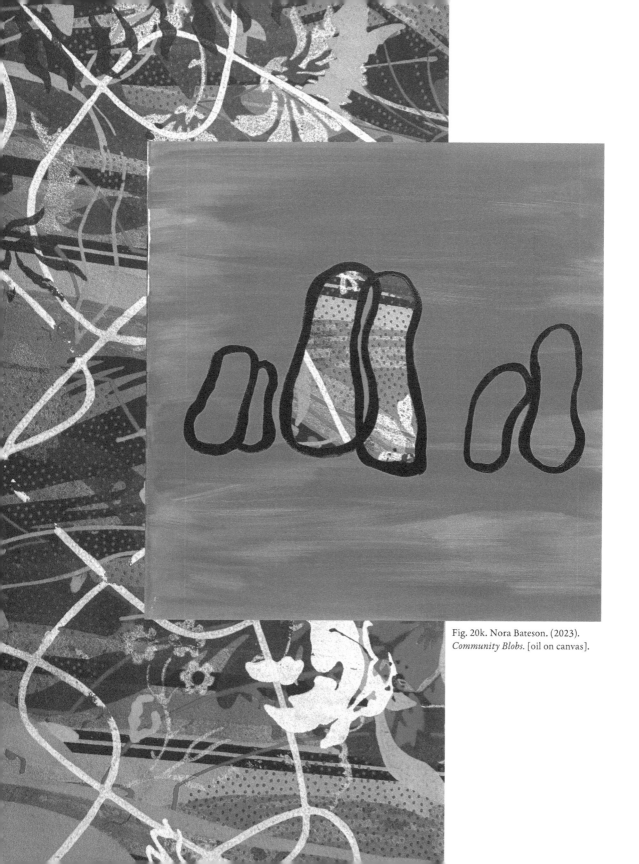

Fig. 20k. Nora Bateson. (2023). *Community Blobs.* [oil on canvas].

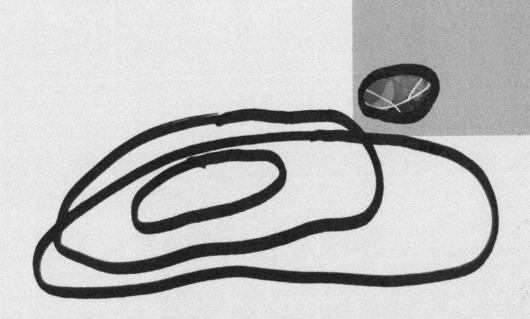

Fig. 20l. Nora Bateson. (2022). Blob History. [ink on paper].

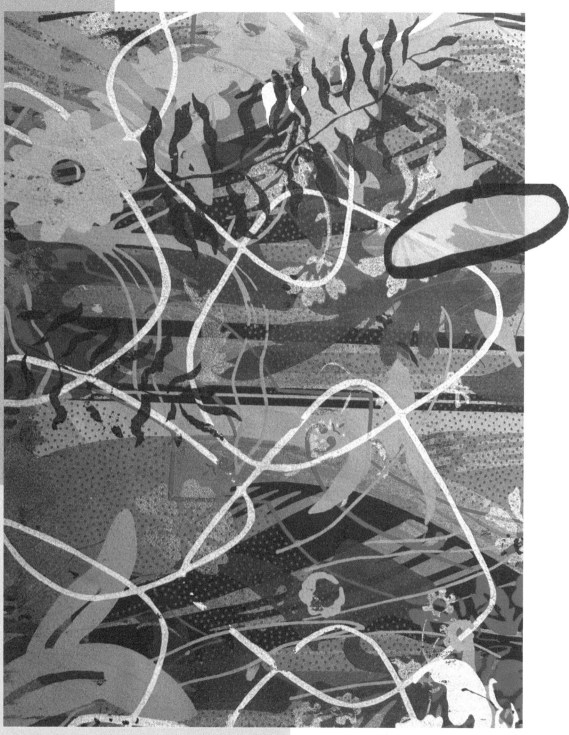

Fig. 20m. Mats Qvarfordt & Trevor Brubeck. (2020–2023). *Untitled*. [Handprinted wallpaper test sheets].

I Fear a Fear of Fear

There is much havoc that comes from fear, no doubt. But there is also something important about fear that must be respected. I am not advocating for fear. I am not so sure it is a good idea to polarize against it, though.

There are things in life that are necessary to be afraid of. A study of mice raised in sealed boxes with no microbial population in their bodies revealed that those mice were utterly reckless. They did not know to be afraid of things that would hurt them. Interestingly, the microbiome is participating in this protectiveness. Is fear in humans any less informed by other organisms that inhabit our bodies? (Arentsen et al., 2015)

How does one know the difference between courage and recklessness? Fearlessness is not necessarily a good thing.

Fear shows care, shows where sensitivity is, shows where the danger is perceived (regardless of whether it is warranted). People fear loss of relationships, loss of respect, loss of health, loss of status, loss of possibility, loss of relevance, loss of coherence in a crazy destructive world.... Fear of being excluded from our "Homo sapiens" groupings is a deep recognition of our interdependency. At its core, this fear is serving the larger project of this species. In other ways, this fear is manipulated and twisted into terrible distortions.

Worse than fear is a lack of sensitivity. Fear has a bad reputation but is also entangled in love, care, and many other feelings.

How fear is expressed, whether it manifests in violence or humility—is another matter.

Perhaps it is not so bad to be afraid. Fear is intrinsic in human complexity. Fear is not possible without love, hope, curiosity, sorrow, or remorse. Fear is time moving forward and backward. Fear is a birthright of being a complex human being, or any being—just as curiosity, care, and sadness are. Perhaps fear has a wider vocabulary in it.

When fear is pulled out from its complexity, held in reductive definition, and decontextualized, then fear is alone, and that is something to be—afraid of?

Cracks and Fissures

Cracks and fissures are breaking the cultural collusion, crumbling the assumptions, confusing the confusion. The way we were taught to think about saving the world was actually breaking it, breaking us. All of human history, some of it horrifying, some of it beautiful—this is our marinating sauce.

Now so seasoned in our constructs and traumas, we have to admit: It was not enough. It is not enough.

The ways of knowing, of loving, of teaching, of healing . . . all of them . . . not enough.

All the religions, the academies, the laboratories, all the art . . . Now, humbly holding a droplet of knowledge in a vessel of curiosity, I have no idea how to fix this world (to avert the course of extinction). And it matters. I only know it matters.

JUST SING

Break the vitriol.

There is no need for all the justifications we might find to polarize.

Let the drama fizzle while we build a fire together.

Or, let's just sing.

FROST

Frost is only doing what it does,
Forming impossibly fantastic crystals,
In asymmetric symmetry,
Fractal fantasy,
But it is only another cold morning,
It's only winter,
Nothing really ...
Just - the breath of December,
How does one simply carry on?
When each small sculpture is an icy feathered forest!
Light sprinkling in sparkled cold,
Finding the scoops and angles of tiny shapes,
Shamelessly,
Every small reach,
Fits into the next,
In chilly unexpected improvisation.
Not perfection,
Just perfection.

TACIT

What is not known
is

What is not said
is

What is not punished
is

What is not written
is

What is not law
is

What is not taught
is

What is not enforced
is

What is not prioritized
is

What is not celebrated
is

What is not structured
is

What is not real
is

What is not repeated
is

What is not on the transcript is
A shift of eye, a catch of
breath,

The architecture of the
conversation
- holds far more information in
its folds.

Is holding –
folded into old habits,

Sneaking obsolescence in between
the lines.

What is breaking,
Is creating.

What is unjust,
Gives justice articulation.

So I won't say the thing –
the saying is empty.

I won't write the thing –
the print is paper.

I will make a new shape of
silences there –
To ping into new connections.

Giving new gaps.
Tacitly forming into new
underground aquifers.

is-ing and was-ing
without is-ing and was-ing.

WILD

De-mapping	=	unclaiming
De-harvesting	=	allowing fertilizing to occur
De-modelling	=	untying from existing establishment
De-strategizing	=	allowing unimagined possibilities

Wild underground unsaid unknown undefined = a possibility zone.

The system cannot change the system.

Combining

Fig. 21. Rachel Hentsch. (2023). *Veil*. [digital art].

What is Submerging?

This is one essay in two pieces, like one cupcake in two bites. In the first piece, I am laying out the enormity and near impossibility of systemic change. Something I hope readers will feel in their bellies like a thud. The perpetuation of the existing ways of thinking cannot address the emergencies of our time; this is not to be underestimated. The second piece is about the unexpected possibilities that await in another order of parallel approaches. These two pieces attempt to express one idea that is not easy to describe without sounding hopeless and abstract, and hopelessly abstract. For me, this is neither. There is a territory of communication, relationship, and daily living that allows for a total change—it is within reach but not in the solutions currently being reached for.

***

There has been much talk of emergence lately. Emergence is a characteristic of complex systems where things happen that set other things into motion. Emergence is a difference that makes a difference between complex and non-complex systems. Emergence makes successful predictions rare and unintended consequences the norm.

Emergence is beautiful and dangerous.

Interestingly there is not so much talk about how emergency is related to emergence... Somehow, in emergencies, discussions of emergence lose their complexity and become pivots of linear solutioning at warp speed.

Meanwhile, I keep wondering, what is *submerging*?

I am curious because it seems something is happening before the emergence. Pre-emergence: conditions for what will emerge are first combining, like soil-enriching nutrients that make it possible for certain plant life to occur. Before the flower is picked, the plant grows; before that, the seed opens; before that, the bacteria, water, and minerals in the soil make mud and create the specific ecological womb in which that particular seed could open. The mixing of many orders of relational processes made the possibilities for what could emerge. Relationships build relationships build relationships build relationships... and *then* after a good deal of responses to responses in multi-directional oscillation, there is emergence.

Combining

If I were to consider the social and cultural pre-emergence in the same way, I might begin to look very differently at what is emerging globally. By the time something has emerged, it is long past the moment of its quickening. The combining conditions in which it was made possible have been stewing into the fertility of distinct possibility for some time, maybe decades, centuries, or epochs. Before something emerges, the ingredients for it to arise have been simmering for a long time. Today's issues did not begin last week or in 2016; they were possibilities fermenting through time. The need to partake in day-to-day life that knits us all into the horrors of industry, warfare, and exploitation—is grown in deep transcontextual processes that have been soaking for generations. The consequences are readily visible in our failing systems and the relationships between them: education, health, economic, political, technological, cultural, and familial.

Out of old scripts and assumptions, entrenched linear thinking habits will seek cause and effect backward and forward in time. Who is to blame? What is the goal? These questions push the possible actions back into the patterns that birthed the problems in the first place. Fixing the symptoms begets more consequences—and around we go. Both questions miss the slow cooking transcontextual submergence. If I were to write a recipe for how to concoct the sort of situations now bubbling up around the world, it might look like this:

> *Cupcakes of Complicit Confusion Recipe*
>
> *Pre-heat culture to broiling competition & individualism.*
>
> *Grease pan with seventy years of banal media entertainment to be sure rigorous thought won't stick.*
>
> *Crack the eggs of mutual respect for each other and the natural world, beat until fragmentation is complete.*
>
> *Add a dash of Military Industrial Complex, Mercer and Murdoch to centuries of colonial violence.*
>
> *Fold in a pinch of Cambridge Analytica (and or) aggregate IQ, simmer until cultural fissures thicken into conspiracies solid enough to sink the real conspiracies to the bottom of the pot.*
>
> *If you can find a pandemic to add to the mix, this can be particularly emulsifying for creating a golden crispy trillionaire crust.*
>
> *Bake for seven hundred years in the oven of mechanistic metaphors.*
>
> *Frosting: Sugar and dreams of material wealth and ownership.*
>
> *Decorate with surrealism.*
>
> *Nothing rises in the cultural cake without something like a garnish of weird.*
>
> *Serve with religion and agriculture.*

From racism to gender inequality to poverty, soil and ocean degradation, and political violence, we cannot fix the systemic issues with explicit, direct correctives—that is not where they live. They inhabit the inflamed scars of previous generations; they grow in the wounds that never healed, they have submerged, and now it hurts everywhere.

In his essay entitled, "From Versailles to Cybernetics," my father, Gregory Bateson, writes, "The fathers have eaten bitter fruit and the children's teeth are set on edge. It's all very well for the fathers, they know what they ate. The children don't know what was eaten" (2000, p. 477). Not knowing what has submerged over time becomes unquestioned attitudes in responding to whatever may happen later. The combining becomes ever more invisible, even though the grooves of the sentiment continue.

> They are living in a crazy universe. From the point of view of the people who started the mess, it's not so crazy; they know what happened and how they got there. But the people down the line, who were not there at the beginning, find themselves living in a crazy universe, and find themselves crazy, precisely because they do not know how they got that way. (G. Bateson, 2000, p. 478)

The bitterness submerged. The trauma submerged. The confusion submerged.

This is why when responding to complex emergent situations, the problem is not the problem, even though it may look like it is. Some try to "multi-solution" the complexity. The itch to pull apart complex systems and list all their components is an impulse informed by old mechanistic thinking, leading to more of the same kinds of problems. You cannot merely fix the parts and reassemble them. That methodology is not going to shift the submergent issues. They will keep reconfiguring. The tending must be to the relationshiping between the parts. And this is messy. It requires perceiving in second or third order, which most people assume is impossible (it's not).

For example, a new curriculum will not change education because education is a consequence of the economy, definitions of success, the job market, and parents' expectations, and all of these will continue to shape preparation for adulthood. Similarly, strict rules will not control tech; they will instead offer challenges for clever go-arounds. Reshaping economy will not change the basic logic of societies asking, "What is in it for me?" No circle, spiral, decentralization, or anything else will change the behaviors that the collective logic of "getting ahead" begets. Those issues are submerged. The potential of "getting ahead" is considered, without saying it out loud, to be a "right"—it is melted into ideas of democracy and woven into the ability to acquire the latest technology or pay for your child's clothing. Success, respectability, love-ability, and credibility are gained by how a person navigated to achieve "independence" in this game.

Meanwhile, there is a brewing mental health crisis. It is even called a "mental health crisis," which speaks volumes about what has been submerged. How should one perform life "sanely" in an insane world?

Combining

Meanwhile . . .

Life continues in some semblance of ever-shifting emergent order. In evolution, organisms change as other organisms change, but some aspects must stay the same, or the organism will die off. But, to stay alive in living systems is also to respond to the constant shiftings inherent in life. So there must also be change. It is a trick of survival to change and not change simultaneously, in just the right ways to be in step with the larger ecology.

As the COVID-19 pandemic rearranged life for so many people, new habits are invisibly settled into the woodwork of daily life. We began to stand further apart. We wore less professional clothing. We witnessed the inability of most of the world's governments to respond to crises effectively. We reconsidered what is essential. These things became the compost of the next round of emergence, most of which remain unseen. Ironically, the seen emergent issues continued to distract attention and generate questions at erroneous levels. Questions like, "Is or isn't online education good for kids?" Ask that question, and a pile of analyses and techniques will arise to increase the value of digitizing an education system that is not serving the students in the first place.

It is easier to look for what is emerging than what is submerging. An emergent situation is easier to perceive—even if that does not give us the information needed. It is perhaps inconvenient to generate inquiry for that which is imperceptible. But things are changing fast. And while the days are filled with news flashes that feel like a tornado of flying cultural flotsam, some significant shifts are taking place and going unnoticed. These are the potent fertilizers of things to come. What is happening now that is quietly slipping under the waterline of conscious observation?

Playing ball with my dog, Blake, I turn in one direction as if to throw the ball that way, and before the ball leaves my hand, he goes galumphing like anything toward where he thinks the ball is going. I tricked him, he noticed, turned with a puzzled look at me, and then full speed tumbled back to look in the other direction. "Which way did it go?" He anticipates, reads my muscular balance and eye movement, and follows the multiple movements that show him my intent to throw the ball—so does a person, family, or society respond to an event with the anticipatory familiarity of prior experience.

The culture of this moment is informed by anticipatory reductionism; it just cannot stop reducing things, even in the vocabulary of "systemic work." Ask a question about economics, and it will be informed by the history of economics, ask about education, and it will be informed by school systems. It is not easy to bring the perception of these processes into the deeper transcontextual submergence that they are truly within. The quest for 1st-order causation is a hard habit to break.

> Living organisms have the equivalent of one "foot" in the past, the other in the future, and the whole system hovers, moment by moment, in the present–always on the move, through time. The truth is that the future represents as powerful a causal force on current behavior as the past does, for all living things. And information, which is often presumed to be a figment of the human mind or at

least unique to the province of human thought and interaction, is actually an integral feature of life, itself–even at the most fundamental level: that of system organization. (Rosen, 2012, p. xi)

How does one attend to the unseen submergent processes? This is a non-trivial question. A system responds as it has been honed to respond. Like a musical instrument crafted through several generations' experiences, the events of our lives find expression through the articulation that our bodies, minds, hearts, and language can find. Prior experience informs future expression. A broken heart makes one more cautious. The past traumas are the filters through which new learning takes place. You cannot return to how you were before the heartbreak or trauma. You have changed, your relationships have changed, you see the world differently, and the world sees you differently. Learning and healing from now on include those experiences. The submerged events are in the soil, and while they cannot be removed, the alchemy, the tone, and the contextual resonance can be shifted.

All of this makes it excruciatingly difficult to imagine what change is needed when the imagination itself draws from the same cupcake recipe as the situation. It is possible, just not in 1st-order solutions. Arguably, systemic change is never achieved with 1st-order solutions. The systemic shifts are taking place in the second, third, and fourth orders and beyond.

I would suggest this is best thought of as murkier than what is known as "confirmation bias"—seeing what you already see—these submerging impressions are baked into our lives within our relationships. They are not rewritten with willpower, agency, or extra effort; they are revealed by getting off-script in relationships. I have been developing a process called Warm Data Labs to meet this need, which is now practiced around the world. I see that most of the next decade's work is in the realm of shifting these deeper perceptions that will allow for another approach. Communities that are fractured, polarized, and fragmented in submergent multi-generational dissonance will not be able to work together. First, a perception shift is required. I am excited about the practice of Warm Data; it shows me daily that there are vast, unfound possibilities for human perception to shift.

After all, in a meadow, each organism is wrapped into life in many ways, not just one, but many. Some are in the songs of love in meadows or the fertile youth of blooming flowers. It is tempting to separate the ecology of meadows from the ecology of ideas . . . but that, again, is just a perception. What kind of relations are made in those ecologies? Who is the earthworm to the bacteria in the soil? To the trees? To the fungi? To the birds? To the grasses and ferns? To the insects? When you think about how each of those organisms knits into relationality with the earthworm—so differently in each case—how those organisms are also wrapping vitality into and with each other, a burst of possibilities is revealed. Their inter-knitting, inter-responding interdependence is continuing in ever-shifting underground submergence. Life-ing.

Next, for this essay, I would like to share a case study of my experience finding old memories in a new voice, with new insight, in a new context, and radically stepping aside old assumptions.

Part 2: Submerging: What it Looks Like

This next piece is a continuation meant to be read in conjunction. I will warn the reader that it is entirely possible to read these pages as a description of an approach to thinking about education. To do so would be to miss the illustration this story provides toward the question: How does one attend to the unseen submergent processes? The context in which this particular breakthrough took place was education and intergenerational learning. But please keep in mind this is possible in all contexts.

In the early days of the global COVID-19 lockdowns of 2020, I wondered if I had something to offer people who would suddenly be home full-time with young children trying to explain the pandemic situation. It was an essential moment to do something for parents wanting to share some of their interest in complexity and systems theory with their kids. I never considered this material too much for little ones because, as I have seen through many classrooms, my own life, and my children—it isn't.

Arguably, kids can do complexity better than their parents. They are less interested in the jargon and present inside the complexity itself. I see this as likely the most essential "education" that the coming generations could ask for. And since the schools were closed anyway and didn't generally offer such a curriculum, I invited parents to join me online for a series on engaging in mutual learning on complexity with their children. My childhood formed the basis of this invitation.

My father was one of the founders of systems theory and cybernetics beginning in the 1930s. I watched him meet each day, meet each interaction, with curiosity and delight in this wondrous complexity. He shared his way of being with me when I was little, and I, in turn, shared something like it with my children. When the pandemic hit, sharing this with others was a perfectly natural response. I had no idea what would happen; I had no plan; I dived in—and that which was lurking in the soil of my being began to sprout. It was as though I had unknowingly waited my whole life for this COVID-19 online class.

Strange. Something interesting happened—something I didn't see coming.

The series started and continued with stories. The first was about what it was like to be a small child in a household where my family constantly questioned the frames through which mainstream culture made sense of the world. I began with the wisdom from my father that we can never be quite clear whether we are referring to the world as it is or to the world as we see it.

As a starting position for parents interested in systems change, my father's thought will keep both parents and kids attentive and humble for potentially several generations to come. We could have stopped right there. "How to Question Absolutely Everything with Your Kids" could have been an alternative title for the sessions. It is an intriguing premise for an education system. Something came loose, an attitude changed tone, and with it, so did the approach toward the whole horizon of learning.

When we began, there was a difference in my approach that I did not pay any mind to. But later, as the sessions progressed, it became clear this was the first time I had entered into a discussion of education—after many years of doing so—in which I was not, in any way, deferring to the existing educational system as a given. The conversations were untethered. I was not hindered by any notions of what education in schools provided. This was the difference that made a difference. There was a wild freedom to actually, really, truly open the field of what mutual learning between generations looks and feels like ... what are the hold-backs? What does it sound like, taste like, to be in discovery of the world with kids? Not *for* them but *with* them ... I was not looking to amend, adjust, or fix the existing education system. This is a critical criterion for getting at another approach to systems change. It turned out that most of the conversation concerned developing a kind of attention to the cultural scripting for parents that places them in a position of non-learning, or worse, "teaching." But in these sessions, the knee-jerk repetition of parental instruction that is so submerged lost its invisibility because we started from another landscape.

> *Do well in school so you can get into a good university.*
> *Read the text and memorize it.*
> *Do well on the test.*
> *Don't spend too much time on the screens.*
> *You have to do your work now so you can be with your friends on the weekend.*
> *One day, you will grow up and have to support your own family.*
> *Success is linear: work hard, make money, get comfort.*

All that stuff ... just sort of crumbled. What graduation? What weekend? What test? What will be the careers that the pandemic will erase forever? What future are the kids headed toward?

The programming of children's relationship to authority, identity, success, economy, and so many other cultural perpetuations was revealed to be generated in the meta-messages between adults—particularly parents and grandparents—and the next generations. The nuance of communication that insinuates and assumes continuation of cultural attitudes and some of the gaps in understanding those attitudes host was remarkably, suddenly talk-about-able. The question was not, "How do I get my child through fifth grade?" but, "How shall we learn together to survive in a changing world?"

I noticed palpable healing in first recognizing and then breaking away from the habituated traps that we, as adults, have subscribed to to survive in a sick system. This became a kind of beautiful punk rock of not foisting those traps onto the next generation—the revolution of simply changing the tone of the parental voice to one of learning. I welled up with passion and declared—No, mothers and fathers ... everything is NOT going to be ok ... unless we get out of this matrix; It is forged between the generations, and there it must be unforged. Like the ring that Frodo has to take to Mordor, where it was made, so too, the generations must set each other free of the double binds of the past. Recognition and rigor are needed to keep a keen eye on normalized, cloaked, inherited habits that form judgments, failures, and unnamed expectations on both parents and children.

These scripts inform life, from school to identity, to relationship with nature, and relationship with ideas of what life is. I think everyone in the sessions sensed how underground and how deeply buried in communication the cultural traps are lurking. This stuff is all submergent. In contrast, the traumas it produces are an ocean of emergent sufferings. The emergencies are ringing all the bells, triggering a response that again taps into the submergent mechanistic logic of 1st-order solutions in the form of pills and programs.

The most profound way to provide one's children with an understanding of complexity is to live a life of learning together in a complex world moment by moment. This means everything changes. Doing dishes is not about "your turn" but about the perception of household relationships. Going into the forest is not about knowing the names of the organisms; it is about wondering how they together are becoming a forest—not knowing turns out to be a position held with far more respect than "being right." The list goes on.

One of the more significant memories that surfaced for me during these sessions was that my father never, ever . . . not even once . . . projected onto me his vision of who I would be in my adult life. He never said, "When you grow up . . ." or "When you go to university . . ." or "When you have children . . ."—to have said those things would be to seed the sort of images that submerge and spin into lifetime notions of success or failure. It took me years to notice what scripts he did NOT write for me. Interesting. The assumption implicit in his NOT saying those things was that I would learn from and find my way through whatever I did. He did not give me instructions, but he had every confidence that I would figure it out as was right for me. I was twelve years old when he died. We knew he was dying, and he knew he would not be at my side when I faced the difficulties of adulthood. So that confidence was a more profound and in-depth, stronger morale than any specific advice he could have given me.

The importance I want to convey here is that the exploration of learning in this format, at this moment, was both an accident and not an accident. It was a stochastic process. Everything I had ever wanted to explore about education, learning, and complexity was unleashed by a change in the context that I did not predict. When school was no longer about school, a whole new world of learning opened. My entire childhood came rolling out from the forest floor of my adult life. I was astounded at the learning I had absorbed so many decades ago as a small child and how easily those memories could be woven into the need for intergenerational learning. Something about the pandemic lockdowns created the perfect soil to tend to an entirely unbound discussion of learning.

That is to say, a different set of assumptions was submerging. And that is what is interesting here. Please don't be distracted by the enticement of a new way of teaching complexity to kids—what I am actually talking about in this essay is the depth of perception change and how most of that is not nameable, not describable, and not see-able, but alters all action entirely.

Education is changed when the communication between generations is in another melody. From the time kids are little, the communication they receive is intoned with messages about their individuation. Failure to individuate is considered to make a person unrespectable, unsuccessful, un-mateable, un-credible. Yet, life is produced through interdependence. What might happen if the assumption was that there would be no individuation, that people need each other, whether they are blood kin or not?

This is a radical notion of a lifetime of mutual learning, mutual tending, mutual getting lost, getting weird, and getting over ourselves.

It is not easy to get out of sensorial loops. Accumulating experience is a way to bank a broad sampling through which to enter unfamiliar circumstances. Collected experiences, since infancy, are further formed through language and describing through relationship to other experiences. Lunch is not just lunch; lunch is a meal, lunch is a meal after breakfast, lunch happens during busy times of day, lunch is the middle, lunch is not with family. Lunch is mixing contexts, just like education is mixing contexts, or economy or . . .

But getting new insight is tricky. Prior experience is tossed into a novel moment like a net sifting out the possibilities in familiar definitions and letting the ones we cannot recognize to fall through the holes. It is similar to when you hear a piece of music for the first time; you find your relationship with the notes and the harmonies through your relationships with other music. But by the seventeenth time you have listened to that piece of music, you will hear and feel it in much more detail. When you taste new food, you find flavors and smells that remind you of what you already know. You are informed by and forming a relationship with the new by tapping into those references that have already been submerged. My father said:

> The *processes* of perception are inaccessible; only the products are conscious and, of course, it is the products that are necessary. The two general facts—first, that I am unconscious of the process of making the images which I consciously see and, second, that in these unconscious processes, I use a whole range of presuppositions which become built into the finished image—are, for me, the beginning of empirical epistemology. (G. Bateson, 2002, p. 29)

The metaphors, the sensations, and the memories may not have precise, conscious identification, but they are there, nonetheless. And they are wrapping new ideas and sensations in old clothes. So be it. Nothing can be done to stop that, which may not be a bad thing. But when participating in any kind of significant learning—BEWARE—these pre-emergence conditions perpetuate and justify old perceptions. What you were able to see before—you will see again. While what you have never seen is harder to see, what has never been said is harder to say.

I am practicing to be a detective of submergence. Without a doubt, this is an intergenerational process. The invisible is not to be underestimated.

Fig. 22. Vivien Leung. (2022). *Play*. [digital art].

Fig. 23. Vivien Leung. (2023). *Moving.* [digital art].

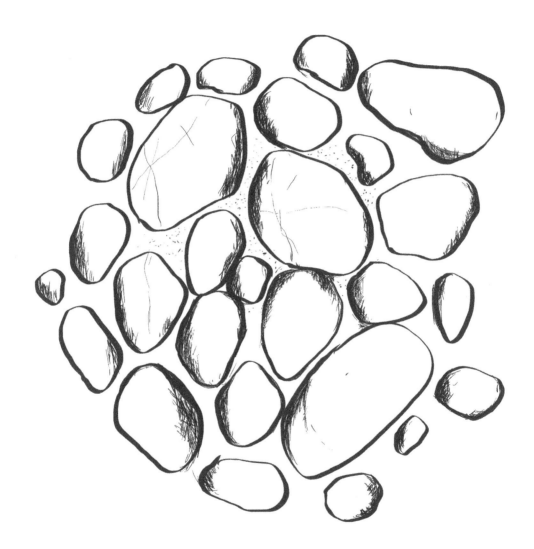

Fig. 24. Vivien Leung. (2023). *Materials Forming*. [digital art].

AFFECTION FOR LIFE

Memories burn and cut and scar, as well as those that heal and give and improvise,
Memories are brewing all the time.

A cosmos in a cosmos.
You have knees,
And clouds above you make shapes,
You are worn by pain and softened in the salty of your tears.

Generations ago things happened so that other things happened so that you and I –
in the outrageous mathematics of life –
are here now.

The possibilities between us are many, most still unspent and unseen.

There is so much we have yet to bring out in one another.

The landscapes we traverse together are always changing – pathways revealing pathways as we go along.

There is so much possibility yet to find ... together.

Affection for life is the air and ground, the pull and the push, the reach beyond our grasp.

URGENT MUD

Inside:
Getting juicier,
The rub of so many paradoxes has worn down the compartment walls of my being,
(Wherever "me" is and however many blurred variables that might include)
Bacteria? Ancestors? An idea, a dance?
Melting forms, grinding down stones.

The grit of disintegrating the boxes gradually got soggy –
Wetness and the goo of tears and the gravy of life's weird turns –
The mud has become a puddle of possibility!
Opaque, formless, and in its own time.
The mud is just sitting there, not doing anything I can see.
But it is doing something.
It is making.
How long it will take to become something? No idea.
How it will begin to quicken and connect? No idea.
How it will shape, what it will be? No idea.
Fusings, musings, combinings are in motion.
In a world of rapid change ...
Fluctuating environments require the ability to respond to unrecognizable circumstances.

Now let it be mud. There is nothing that will come otherwise.
This is urgent mud.
It ferments,
starts something,
a small unseen stickiness, then more –
Then there is a happening.

Outside:
I am still pinched by the collective exoskeleton of ideas on how to fix the nice muck,
Name it, plan it, shape it,
The wrongness of it –
The uselessness of it –
The pointlessness of it –
The purposelessness of it –
Oh yes. It must be thus.
Precisely unsee-able from here.

Later, maybe there will be a moment to say, "Ah ... it was there all along."
Will saying this negate the worlds of creative work in the precious dark mud it took to allow new forms through?
Will we forget that it took being lost?

UNTAMED

Senses
Hardly met,
New languages hardly spoken,
The descriptions still mostly unmade,
The textures unsmelled,
The unseen submerging is a world of —
Gaps,
Not empty,
But,
So full of possible possibles.

There is no time to despair,
The discovery has hardly begun.

Aphanipoiesis

Building upon two other neologisms, "Warm Data" (N. Bateson, 2017) and "Symmathesy" (N. Bateson, 2016), a third has become necessary in order to describe an unseen property within living systems. I introduce the word aphanipoiesis—meaning a coalescing of unseen factors toward vitality.

This chapter is an introduction, primarily to the new word, and will describe its place with the other two words as observed within the Warm Data Lab research over the years. All three words have their basis in abductive process and explore the inevitable mutuality of formation through a transcontextual relational process. As an initial exploration into the aspect of change occurring in living systems characterized as being "unseen," this inquiry is a beginning point and invitation from which the exploration can be deepened. The criteria and formalities of this process remain unnamed and undefined. The following are some preliminary observations and questions to be pursued together.

Gregory Bateson, my father, is referenced repeatedly here as this new theory is rooted in his work. In my studies of systemic practices and explorations of complexity theory, I have often sensed an essential underlying understanding of living systems in my father's work that is difficult to point to but somehow sets his work apart. My father carefully wrote an awe for how life keeps life-ing into his texts. Not as a dogmatic mystery, but as scientific rigor, he sculpted his descriptions of systemic process so as never to nick the artery of indescribable, infinitely entangled communication in families, forests, and societies. These careful wordings and observations are easy to miss because they are not spelled out as such, but they are there. If one reads Gregory Bateson for solutions and methodologies, it will not take long to become exasperated and toss his work aside. It is not convenient or easily applicable; this way of seeing eschews the urge to dive into action in explicit ways.

To attend to the unseen requires patience, hard work, and seemingly endless searching through infinite lenses. By no means is this exploration a suggestion of "surrender" or "acceptance" in new age terms. On the contrary, this is an invitation to add to the existing studies of living systems those critical characteristics that make it possible for organisms and societies to harbor potential change in unseen ways long before the new policy is adopted or the new limb begins to grow.

The change before the change suggests that perhaps indescribability is, in itself, an evolutionary condition, a built-in extra budget of possibility for unfamiliar formations. I suspect it is, and this indescribability offers the challenge of how to communicate this unseen, submerged process. One angle is to accentuate the lateness of responding to "emergence"—when what has emerged is the expression of stored impressions that have long since been gathered. What about pre-emergence?

In linear causal terms, it is not difficult to point to an emergent situation and suggest its opposite as a response. Dry soil requires water, hungry people need food, and high levels of carbon in the atmosphere need to be removed. Yet complex systems defy this sort of 1st-order response with unintended consequences; most of today's most significant "problems" result from yesterday's "solutions." One cannot assess a person's health only based on diet and determine that they would achieve ultimate health if they ate a different diet. This is not so. The entire relationship with food reaches into geography, culture, early childhood, stress, old injury, both physical and emotional, and countless other contributors. The same can be said of more significant issues like climate change; the problem may be carbon particles in the air, but the long history of cultural respect being produced through material wealth has guided the behavior of rich and poor communities alike, fueling the destructive and exploitative industrial production-distribution-waste cycles. The issues are formed upstream of the emergence in the earlier overlapping and combining of unexpected and subtle experiences. Those subtle day-to-day subterranean learnings form and continue to form each of us.

This unseen realm is vital, non-trivial, and sacred, and it is real. I am increasingly finding that the most fecund realms of change, learning, and evolution are beyond the organism's current capacity to perceive. The flexibility that lurks below conscious perception is like the soil beneath the forest, teeming with relational processes. While most attention is caught up in what can be perceived, there is a wildness in the implicit correlations, connections, and coalescing impressions.

I am deliberately making a correlation between these unseen accumulating stories in biology and human society and suggesting other ways of thinking about systemic change in human society. In this era, the culture of change-making seems to have an unfulfilled attraction for "systemic transformation" and "emergence." This tells us something about how systemic change is perceived. The filters of perception aligned only to strategic action obscure essential information about how submergent, inherent, unrealized information is forming. It is easier to name what is seen and strive to change it. But what if what is seen is already old news? Is the search for direct correctives a search in vain and looking in the wrong places?

Like the story of the man looking for his keys at night—only where he can already see under the streetlamp— it is time to ask: "How to stop looking for the car keys under the streetlamp when it is known the keys were lost in the forest?" Just because it is possible to measure and describe emergent events after the submergent coalescence does not justify turning away from the difficulty of addressing their nascent becoming. These pre-emergent processes are more challenging to define but at least equally necessary to consider.

Many will find this a frustrating inquiry, especially in this era of urgency in climate, culture, and economic danger, often referred to as "the Anthropocene." It is possible some will call out this study as a frivolous fixation on the subtleties and nuance of perception. In this writing process, I have felt at times like the emperor's tailor, vying for the rich beauty of invisible silks. However, explicit action plans and impact reports have been unable to touch the realms of deeper impulses and multi-storied impressions from day-to-day decisions. The crises of our times can primarily be described as insidious. A tiny gesture toward healing implicit, destructive relational habits is much more practical than a hundred attempted explicit action plans that result in caustic erosion of more relational connective tissue. More than that, it seems essential to discover how evolution requires a shared capacity to meet an unknown future: The ideas in civilization are (like all other variables) interlinked, partly by some sort of psycho-logic and partly by consensus about the quasi-concrete effects of action.

> It is characteristic of this complex network of determination of ideas (and actions) that particular links in the net are often weak but that any given idea or action is subject to multiple determination by many interwoven strands. We turn off the light when we go to bed, influenced partly by the economics of scarcity, partly by premises of transference, partly by ideas of privacy, partly to reduce sensory input, etc.
>
> This multiple determination is characteristic of all biological fields. Characteristically, every feature of the anatomy of an animal or plant and every detail of behavior is determined by a multitude of interacting factors at both the genetic and physiological levels: and, correspondingly, the processes of any ongoing ecosystem are the outcome of multiple determination. (G. Bateson, 2000, p. 508)

Insidious

I have attempted to name and discuss a characteristic of life I have found necessary for any systemic change. Most of the urgent issues at this moment in history can be described as "insidious," which is to say that they are produced through the combination of circumstances over time in unseen ways that have produced danger. Racism is insidious; sexism is insidious; corruption is insidious; consumerism is insidious; greed is insidious; cancer is insidious; trauma is insidious; addiction is insidious. A contemporary definition of the term "insidious" is roughly "spreading gradually or without being noticed but causing serious harm" (Oxford University Press, n.d.) or "(of something unpleasant or dangerous) gradually and secretly causing harm" (Cambridge Dictionary Press, n.d.). Note: The older definitions refer more to ambush or to lie in wait, which are closer to the etymology of the word's more literal definition: "to sit in."

But how would one describe the opposite of this?

What is the word to describe how unseen, gradual processes come together to form life, vitality, healing, and ongoing learning?

And how would one know where the vitality begins, and danger ends?

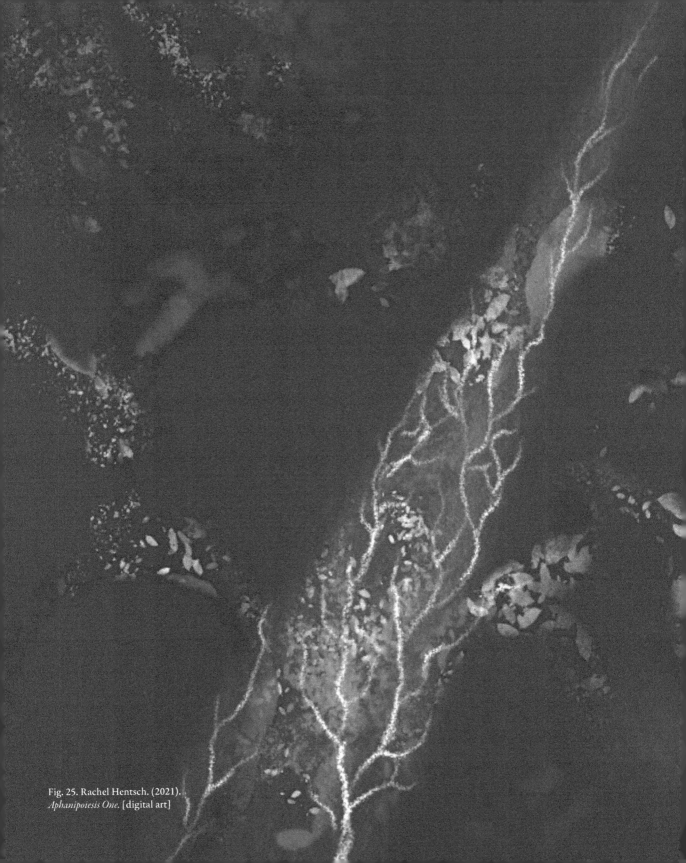

Fig. 25. Rachel Hentsch. (2021). *Aphanipoiesis One*. [digital art]

Living systems are a constant combining of multiple forms of communication and interaction between organisms. While it may be possible to point to some of the 1st-order combining and communications in living systems, the second and higher orders of communication remain unseen, inseparable, undefinable, and crucial to the trajectories and aesthetics of ongoing vitality. This process of unseen coalescence could use a name. By bringing together two words from ancient Greek, I propose the word aphanipoiesis as a term for the way in which life forms in unseen ways:

Aphanis: Greek root meaning obscured, unseen, unnoticed.
Poiesis: Greek root meaning to bring forth, to make.

A possible definition of aphanipoiesis could be:
Noun (n.) An unseen coalescence toward vitality.
Noun (n.) A coalescence of experience becoming unseen.
Other grammatical forms could include the adjective aphanipoietic (adj.).

Other words that also carry the root aphanis include "phantom," "diaphanous," and "phenomenon," while the root poiesis is familiar from the word "poetry." Afanis has the implied meaning of "something standing back, in the shadows, in humble quietude," and poiesis is seen in the familiar work of Humberto Maturana and Francisco Varela (1980) on "autopoiesis." The coming together of these root words is not simply a sticking together of the two ancient words but also brings together the history and contextual meanings they have gathered along the way. The making of the word aphanipoiesis has, in itself, held unseen assumptions and habits. The meaning of the word rests in an ambiguity between making the seen into the unseen and generating vitality through the unseen. Both are appropriate. Even though I may have been conscious and "seen" various experiences in my life, I did not "see" how they coalesced into other experiences to form the premises of my other thoughts. So, both the unseen coalescing toward vitality and the coalescing making experience unseen are taking place. As an aspect of life life-ing, aphanipoiesis is neither good nor bad but neutral. Vitality can be dangerous, too. Sickness can catalyze immune systems, and forest fires are needed for certain trees to grow.

It is a reprieve from the structure and control of multi-cause and effect models to recognize that perhaps life requires something like a diaphanous poetry that rests in organisms as they meet an ever-changing future. It is difficult to measure such a thing, which may bring a different approach to the study of change. Undoubtedly, the possibility of aphanipoiesis as a characteristic of living systems will also elicit an ocean of questions.

Of note is the importance of this coalescence being "unseen" rather than "hidden" or "invisible." There is nothing hidden in this natural process. It is merely out of habituated perception. When I first hear a piece of music, say a flamenco guitar piece, I may respond to the harmonies and instrumentations. But it may not be until I have listened to the piece several times, or until I have studied guitar, or studied Spanish, or the history of the oppression the song tells of that I am able to consciously perceive the hints and gradations in the notes and rhythms. They are not hidden. They are unseen. Also of note is that the term "unseen" is being used here to refer to other senses as well: unheard, unsmelled, unfelt, and so on. Older cultures have held the unseen to be relevant. It will be fascinating to explore how aphanipoiesis connects to those older epistemologies:

> Nothing happens here that did not begin in that unseen world. . . . The indigenous understanding is that the material and physical problems that a person [institution] encounters are important only because they are an energetic message sent to this visible world. Therefore, people go to that unseen energetic place to try and repair whatever damage or disturbances are being done there, knowing that if things are healed there, things will be healed here. (Somé, 1999, p. 23)

There is a definitive difference between this study of aphanipoiesis and the study of the unconscious. What is being explored in this instance is not "that which is not seen" as such—but rather the coalescing of many experiences that may have once been perceivable. It is the merging, mixing, and fusing (aphanipoiesis) process that is being featured through this study. The characteristic that differentiates this from other work on the unconscious is the attention to coalescence.

STOCHASTIC FRACTAL FLEXIBILITY

Evolution does not know where it is going. Or rather, to get where it's going, which is to the continuation of life, the world of living systems is built, ready to find a way. We are on the bus with a planet full of organisms, all humming and buzzing with life moving through time, shifting, learning, changing, and responding to each other, but there is no destination. Predetermined goals and destinations are efficient in industry but inefficient (even incapable) in continuing the complexity of life. The goal of continuing life is another level of "goal" worthy of a planet full of wiggling organisms. If the larger "goal" is to preserve the possibility of ongoing complexity, the most vital trait we collectively participate in is the conservation of flexibility. The movements of each organism—our shapes, our rhythms of communication, our energy sources (food), our excrements, our birth, our death—all are in constant improvisational change, allowing the premise of life in the form of communities, societies, oceans, forests, meadows, soil biomes, and families to continue.

All of us, including the starfish, the redwoods, and the trillions of organisms that live in and on my body, remain in an intergenerational production of the "multiple possible." Living organisms have to stay in relationships, and these relationships must be able to change, and the changes must remain wild. To decrease the complexity is to increase the likelihood of loss of flexibility or obsolescence. Perhaps this flexibility is produced through aphanipoiesis as the unseen coalescing of implicit communications in living systems—brews sunken, latent.

Stuart Kauffman (1995) points to the other side of this phenomenon when he describes the possibilities that open with each incidence of emergence. The tiny and huge events that once occurred create the context for other events. The invention of the cell phone made texting and a whole culture of emoji culture possible, for example. Kauffman calls this the "adjacent possible."

Fig. 26. Rachel Hentsch. (2021). *Aphanipoiesis Two*. [digital art]

How does a living system change?

Ultimately, this question is rooted in the inquiry into how evolution takes place. Ongoing life is a combination of both chaos and order. Order persists but changes if there is stored flexibility in the system to allow it. Nearly random, but not entirely random events—that is, stochastic processes—open the pathways winding toward change. A big storm, a migration, a new fungus, loss of species—day after day, century after century—life brings surprises that all organisms have to tuck into their memory. Recessive genes remain in storage, uncalled for until there is a contextual beckoning for a new combination.

There is a paradox in the processes of how organisms evolve: the process of changing requires continuance; otherwise, the changes lead to non-viability or obsolescence rather than continued relevance or interdependence. Adding to this paradox is the impossibility of explicitly tracking those changes when they are multi-faceted responses to relationships taking place at many orders of ecological vitality. They leave traces to be felt later when needed, but we do not get to say when that will be. The stochastic process of shifts in response to shifts through multiple contexts is so boggling that it interrupts the hunt for "cause" and pushes inquiry to another vista. From this vista, the possibility of aphanipoiesis becomes beneficial in opening up new forms of exploration of the cache of flexibility that allows for such unforeseeable "multiple possible" responding. Where does the "new" come from?

Somehow, a meadow, a forest, or a family continues to exist while changing at every order—until the relational tapestry that holds the forest into its continued forest-ing becomes too threadbare. Then, the obsolescence (devitalization) occurs. One could examine a family through their entanglements into the contexts of intergenerational relationship within cultural, economic, and political dependencies. It is also critical to recognize that the family wound is utterly contingent upon the microbiome and soil for food, air, water, wood (forest), metals, and so on. As culture changes, the assumptions around the notion of "family" are shifting. Still, it is impossible to venture a guess into how economics, politics, education systems, health systems, religion, changing microbiome, climate, and so on are impressing upon these new forms. That they are changing is indisputable, but the realms and ways they are changing remain stochastic in their unfolding. Where is the edge of the family? Where is the edge of the forest? Does evolution require something like a process of aphanipoiesis?

> I was laying down very elementary ideas about epistemology—how we can know anything. In the pronoun *we*, I, of course, included the starfish and the redwood forest, the segmenting egg, and the Senate of the United States.
>
> And in the *anything* which these creatures variously know, I included "how to grow into five-way symmetry," "how to survive a forest fire," "how to grow and still stay the same shape," "how to learn," "how to write a constitution," "how to invent and drive a car," "how to count to seven," and so on. Marvelous creatures with almost miraculous knowledge and skills.

Above all, I included "how to evolve," because it seemed to me that both evolution and learning must fit the same formal regularities or so-called laws. (G. Bateson, 2000, p. 4)

TRANSCONTEXTUAL

"Transcontextual" is a descriptive term I discovered in Gregory Bateson's *Steps to an Ecology of Mind* (2000). I have found it more valuable than I imagined. Attention to what is happening across and between contexts moves the study of change from change in the organisms or parts of a system to change between them. I am less interested in "transcontextuality" in noun form, which loses the relational aspect of the adjective. This idea proved to be more significant than I expected. The following quote addresses the breadth of the concept nicely. In this passage, my father refers to the syndrome of the double bind, which is a bind that has caught a person or an organism in a situation in which they cannot succeed without a jump in perception:

> Let me coin the word "transcontextual" as a general term for this genus of syndromes.
>
> It seems that both those whose life is enriched by transcontextual gifts and those who are impoverished by transcontextual confusions are alike in one respect: for them there is always or often a "double take." A falling leaf, the greeting of a friend, or a "primrose by the river's brim" is not "just that and nothing more." Exogenous experience may be framed in the contexts of dream, and internal thought may be projected into the contexts of the external world. And so on. For all this, we seek a partial explanation in learning and experience. (G. Bateson, 2000, p. 272)

I regularly use the following quote in the Warm Data Lab training to illustrate how an experience is received and perceived through many contexts simultaneously and without conscious noticing. "It's not just that and nothing more" (G. Bateson, 2000, p. 277).

Warm Data is defined as transcontextual information produced through communication forms, both direct and indirect, that combine contexts. A friend who waves from across the street may be waving for the first time since an argument; they may be waving instead of crossing the street to embrace you; they may not be waving at you but at the person you are walking with. The meaning of the wave is a blend of contextual stories. The blended contextual overlapping is particular to each circumstance, social or biological, but the blending is universal.

Whatever it is—it is not just that and nothing more. The details and specifics that are blending produce an aggregate of variables for possible change or flexibility, but that aggregate can also create double binds, stuck-ness, and other traps. The point is these many-tentacled blurring stories are holding both continuance and discontinuance. To evolve (which may include a period of catastrophe during reorganizing) or to become obsolete?

These issues are responses to responses. The need to create a respectable and perhaps attractive identity in a society that reveres material wealth makes consumerism and monetary ambition a response to the more ancient need to be relevant to one's community. Counting carbon particles and producing media to communicate climate change are responses to these responses. They are all so deeply transcontextual and combined in an alchemy of subtle and not-so-subtle transcontextual affirmations that they are not successfully "fixed" or removed by explicitly formed direct correctives. In fact, they are usually made more dangerous by such responses.

A direct corrective to an insidious issue is most often experienced (in human terms) as confusion, rejection, or humiliation. The "problem" is "not just that and nothing more"—forming between contexts in unseen ways. Consumerism is implicit in language, education, economy, toys, family relations, law, and more—to try and "fix" it, as an issue to itself will strike multiple contexts of response, or in other words, it will generate all sorts of confusion. Yet somehow, a meadow knows how to evolve, both in its organisms and in the relationships between them. Something is continuing—the meadowness—and much that is not continuing both in the organisms and between them.

Warm Data

As the International Bateson Institute began to research "How Systems Learn," it became clear that there was a need to describe another kind of information that could hold the ever-shifting and responding of transcontextual processes. This description became "Warm Data" (N. Bateson, 2017). Warm Data was coined in 2012 to describe transcontextual, relational information. Warm Data is information that is alive and shifts within the mutual learning of all living systems. Information that describes the ever-changing relationships between contexts must hold 2nd-order cybernetics, paradox, and authorize perception of the aesthetic, texture, or tone of the meta-communication in the systems. The practice and development of the Warm Data processes brought a strong taste of the potentialities in this lively zone of contextual overlapping and reframing. But it was impossible to pin down, impossible to predict, and impossible to see.

Symmathesy

First coined and published in 2015 and presented at the International Society for the Systems Sciences (ISSS) meeting in Berlin of that year, "symmathesy" is the study of how systems learn through transcontextual mutual learning (N. Bateson, 2016). This study has opened up many new prospects of inquiry through the question, "How is it learning to be in its world?"—a crooked tree, a disturbed child, an insect in the jungle.

The inquiry that emerges from this symmathesetic question pushes the observer into a perception of how the studied system is described through its relational contexts. The tree is learning to be in the soil it is in; it is learning to be in the shadows of the trees nearby; it is learning to be in relation to other organisms that are, in turn, learning to be with the tree. Symmathesy, as a concept, has been brought to evolutionary biology, family therapy, healing conflict, political polarity in communities, and more.

ABDUCTIVE PROCESS, TWO VERSIONS

As a process of coalescing, the new term "aphanipoiesis" asks for an approach to the study of living systems based on how organisms and ideas alter each other, mutually learning, simultaneously allowing for stasis and change. If all organisms moved through time and stayed the same, like a machine, life could not continue. Shifting is relational and responsive. The study of this responsiveness is inexhaustible and eternal. The study drops out of concretized definition and into another form of inquiry that requires casting communication across contexts and considering the transformations in that movement.

Charles Sanders Peirce (1998) and Gregory Bateson (2002) recognized the importance of abductive process as necessary to obtain new information or insight. Peirce was more focused on how the information from one context could provide the basis for the hypothesis in another context. Bateson was interested in how one context in a living system became a description of the others. Both of these approaches to abductive process are viable in the study of aphanipoiesis. Peirce articulated the three forms of logical meaning-making: induction, deduction, and abduction. He went on to note that abduction has the importance of offering the possibility of new ideas:

> Abduction is the process of forming an explanatory hypothesis. It is the only logical operation which introduces any new idea; for induction does nothing but determine a value, and deduction merely evolves the necessary consequences of a pure hypothesis. (Peirce, 1998, p. 216)

I am both surprised and not surprised at how little the current discourse of systems transformation has drawn from the theory of abductive process—surprised because it is critical to understanding how systems change, not surprised because it shifts attention from "fixing" a system by adjusting its parts to attending to the possible relational processes within that system and between that system and its "environment/s." The latter is un-measurable, and manipulation of it is dangerous.

Gregory Bateson provides language around the nature of "description"—while Peirce opens the door to consider the process of lateral pattern recognition. Both scholars recognize the importance of abductive process in thinking about change in living systems.

> All thought would be impossible in a universe in which abduction was not expectable. Here I am concerned only with that aspect of the universal fact of abduction, which is relevant to the order of change that is the subject of this chapter. I am concerned with the changes in basic epistemology, character, self, and so on. Any change in our epistemology will involve shifting our whole system of abductions. We must pass through the threat of that chaos where thought becomes impossible. Every abduction may be seen as a double or multiple description of some object or event or sequence. (G. Bateson, 2002, p. 134)

Aphanipoietic Realm—Where Flexibility Lives

In the moment of response to an event, the organism evaluates its circumstance as it can, using its history to inform its future. Evaluation of context is as vital as water. The question becomes: "How is the information of past experience organized into the evaluation?" This is not a direct corollary. As an example, I learned in math class to use formulas that were not necessarily useful to my life in terms of working through algebraic equations. I hardly ever use my knowledge of algebra as such, but as any mother can attest, each day is an ongoing calculation of multiple variables of time, temperament, sleep, food, finance, and tradition. Learning to think in terms of variables was indeed useful.

In algebra class, I also learned how to conduct myself in relationships with authority; the establishment of successful or failed communication with the teacher later informs relationships with employers, in-laws, etc. This learning was affirmed and confirmed in a myriad of other contexts, like getting a driving license or going to the doctor, where attention to cultural hierarchical deference is required for success. Abductive learning is constantly bringing the experience from one context into another, informing the well of named and unnamed experiences from which to draw instant evaluation of emergent circumstances.

In my work, the snag I have with abductive process is that it pushes the strategist and the committee into chaos. For example, if one asks about a crisis in the education system from an abductive stance, the question defies all familiar paths of solution. Where is the problem in the education system? Is it in the classroom, the family, the economic system, the communication between generations? Is it cultural? Is it historic? Is it technological?

Through the Batesonian lens, the education system can be seen as a "description" of these other contexts. Investigation of the transcontextual descriptive process begins to illustrate how the education system is a description of the economy and job market into which students are funneled. It illustrates how the education system is a description of intergenerational expectations—the education system is a description of a culture that values reductionism; it is a description of the history of education. In each case, the contexts are inter-descriptive, forming abductive zones of relational, semiotic processes. The snag is that no innovative curriculum can undo the knot of overlapping, inter-stitched contexts that produce what we know as an education system.

So, where is the change? Usually, this question is posed toward a perceived need for action toward a predetermined goal, effectively stomping on the possible aphanipoietic potentialities and leaving insidious habits to suggest direct correctives, only later to be repelled. The way our parents' expectations inform our understanding of how to create an identity, how we wind that into our ability to get a good grade in algebra or a titled position at a firm, how these expectations melt into what it means to be a citizen; and to be a parent one day ourselves—all of this is an abductive transcontextual swirl of responsiveness.

Aphanipoiesis is what happens in this messy but vital commixing of responses. The diversity of impressions that are changing each other can be compared to an ecological system. The pathology of

Fig. 27. Vivien Leung. (2023). *Mama Hands*. [digital art].

the education system requires this diversity, not just in the number of perceptions but also in the strange blends that can release a new perception. At the same time, the possibility for any new forms to come into being must surely start in this aphanipoietic process. The necessity of it being imperceptible is that it is kept outside of the habituated forms of action that reconfirm each other—a form of flexibility of the utmost value to an ever-unfolding evolutionary story.

> Social flexibility is a resource as precious as oil or titanium and must be budgeted in appropriates, to be spent (like fat) upon needed change. Broadly, since the "eating up" of flexibility is due to the regenerative (i.e., escalating) subsystems within the civilization it is, in the end, these that must be controlled ... flexibility is to specialization as entropy is to negentropy. Flexibility may be defined as uncommitted potentiality for change. (G. Bateson, 2000, p. 505)

The Study of Change

The study of change is usually caught in the cultural cul-de-sac of measured differences in decontextualized outcomes; a more relational study would note that changes that have become visible or measurable are not the change itself but rather the consequence of more interrelational shiftings within a living system. Therefore, it is helpful to consider the abductive processes at work in the deeper formation of relational change. The ways in which the multiple entities of a living system are continually responding to the shiftings of each other are moving potentialities for change. It may be possible to name the changes once they form, but by then, the deeper abductive possibilities have long since been brewing across and through multiple contexts. The need to find language through which to approach a discussion of this pre-emergence realm of communication prompted the forming of the new word to describe it.

I have previously referred to it as "submergence" in an earlier chapter, but I acknowledge that language is probably inadequate to address the complexity of the process that aphanipoiesis can hold. While there may be a kind of submergence of unseen impressions that produces a basis for the pathways of what will later appear as emergence, the submergence alone is not stirring the pot. To stir the pot of existing underpinnings of sense-making, combining experiences through multiple contexts sets things in motion. A meadow is only a meadow through the many forms of communication and relationship between the organisms. The earthworm is in different "mutual relationshipping" with the soil, the trees, the grasses, the insects, and so on. Each of these organisms is correspondingly in multiple relational processes with the other organisms of the meadow; the vitality of the meadow is continued through these many responses reflecting through many contexts.

Hypothesis and Anticipatory Systems

For Charles Sanders Peirce (1998), the abductive process offers an opening of understanding through hypothesis, in which new connections can be made. In keeping with the example above, we might hypothesize that understanding contextual patterns of the education system can be projected onto the economic system to understand both better. Right away, it is evident that this transcontextual pursuit unveils similarities and differences that give insights. The beauty of this process is that it opens up new ways of understanding, redoubled when many observers are multiple hypothesizing.

As the hypothesis is forming, there is also the possibility of noticing what latent presuppositions are laced in and how those unseen epistemological preconceptions are setting the limits of the hypotheses and nudging directions of response. The hypothesis itself has been informed by a history of experience, collected into an unseen set of favored understandings. If I know how to find my way around an airport, I will be able to use that familiarity in other public buildings. I will look through that familiar experience into the one I am faced with and source from the familiar to apprehend the unfamiliar. If I am considering starting a partnership, I source information from my experience in other relationships (with parents, lovers, siblings, and friends) and business. These favored understandings might be insidious, edging toward danger, and they might be aphanipoietic, edging toward vitality. It is necessary to ask, what are my hypotheses telling me about the pre-existing patterns of thought—influenced by culture, education, and economy—that filter and select the form of my hypotheses?

Applying the idea of an aphanipoietic history to hypothesis gives another layer of information that is humbling to the eager strategist. It is interesting to note that the way in which the above hypothesis is formed is, in itself, an illustration of many of the pre-existing epistemological sensitivities formed gradually through the coalescence of many other contextual impressions. I have an already existing lens through which I am tracking ideas familiar to me. The hypothesis reveals these familiar perceptions and abilities to perceive and will lead to entirely different directions of decisions or actions depending upon habituation. The hypothesis itself becomes a tender risk or vulnerability, a permeable moment when the limits of perceptive capacity are revealed. Exposing existing familiarity with something in one context so that it might shed light on another is a basis of the experiencing of any newness. A new flavor is explored through the experience of known flavors; a new form of music is explored through understanding other forms; thereby, the abductive process becomes a zone of untamed, unnamed, unseen, and essential contributors to what may later be called emergence. But one person may taste a spicy dish and recognize the spice as discomfort, while another may recognize it as an artistic culinary skill.

The hypothesis is not bland. It is not uninformed. Instead, it expresses the deeply held, habituated assumptions that are ready to receive new combinations of contextual information. In this way, perhaps the more insidious premises are brewed at the level of aphanipoietic process. The abuses experienced by children later fuse into other experiences and set the limits and the tone of hypothesis in other relationships. The early experiences may be long forgotten, but the pain still informs in ways that are not easily adjusted at conscious verbal levels. Likewise, many experiences of success or struggle in the classroom may inform later responses to employers and authority figures.

Robert Rosen's (2012) work on anticipatory systems cautions that when considering the responses of an organism in a system, it is always necessary to look into the historical sensitivities of (or within) that organism that produce particular proclivities for particular aspects of what is otherwise a broad spectrum of information. The organism picks up on what it is anticipating, what it is already familiar with, what we have words for (in human experience), and what is already known. In the preface, Robert's daughter, Judith Rosen (2012, pp. xii-xiii), writes:

> All "instinctive" behavior of living organisms is based on the activity of such internal predictive models, generated from encoded information within their own systemic organization. To observe and learn about the annual migration of Monarch butterflies in North America gives us enough evidence to put us in awe of just how detailed the encoded information can be and how powerful the guiding action is of these internal models on the behavior patterns of all living things.
>
> There are stark dangers embodied in this situation, however . . . The dangers stem from the fact that many of the encoded models (and/or the information from which they are constructed) are either not able to be changed within a single organism's lifetime or else they change too slowly to be able to avoid disaster in a rapidly changing world. There is no way to know, from within a model, that the system it was encoded from has changed radically. The model will keep on making predictions using wrong information—and the organism will still be guided, partly or entirely, by those predictions. If the predictions are inappropriate, the behavior will similarly be inappropriate—perhaps to the point of mismatches that prove lethal to the organism. Because organism species within an ecosystem are so interlinked in their requirements and dependencies, the death of significant numbers of one species can initiate further rapid changes in the behavior of the local environment, which can ultimately cause rapidly escalating cascades of extinctions.
>
> This is the Achilles heel, the innate vulnerability of all anticipatory systems. With human-induced changes to the composition of Earth's atmosphere happening at an unprecedented pace over the past two hundred years, and the further unknown changes which are likely to be caused by them, we would do well to pay very careful attention to the warning that is inherent in these facts. Any model-based guidance system will only be as good as the encoded information it uses.

In *Mind and Nature* (2002), my father provides six criteria of what he calls "Mind." These criteria are not about brains but are instead a set of possible laws of living systems whose interdependent aliveness he calls Mind. The fifth of these criteria is essential to the idea of aphanipoiesis because it describes how an organism receives any input into an already grooved set of experiences (differences). Criterion Five of Mind: "In mental process the effects of difference are to be regarded as transforms (that is, coded versions) of the difference which preceded them" (G. Bateson, 2002, p. 102). To get to the heart of this idea, it is necessary to diverge for a moment into what Gregory Bateson means by "differences." The way that he defines information is as the "difference that makes a difference" (G. Bateson, 2002, p. 212). Comparisons between sounds, colors, textures, tastes, distances, weights, emotions, tones, and so on provide an ever-increasing collection of contrasts that allows for refining perception. The way musical notes are played together, and next to moments of silence, is information about the musical notes as single notes and combinations. A single note played forever actually is nothing except in contrast to the other sounds of silence or your heartbeat. The "difference that makes a difference" is a relational description of information that contains the idea that comparison and distinction will be particular to the organism receiving the information. I may hear a piece of classical music differently than my father or my children do because the harmonies will call up experiences of my life (musical and non-musical), producing a unique set of "differences that make differences" in me. I taste my daughter's lemonade differently than you might because you have a different history of lemonades, daughters, drinks, summertime, mothers, kitchen smells—and so on.

When Gregory Bateson (2002, p. 102) says, "In mental process, the effects of difference are to be regarded as transforms (i.e., that is, coded versions) of the difference which preceded them," essentially, he is saying that as new experience or information is met, the organism, or community of organisms, will receive that experience by referencing those familiar experiences. I walk in the Swedish forest and think, "This is like Connecticut." You might taste a fruit you have never had before, and something in you is searching for recognition—it is like an apple, or no, a mango. I don't know. But the point is this—the new experience is not met with a blank slate. The organism cannot discern the new experience's nuance, or even its more basic information, without attaching it first to some kind of reference formed from previous experience. The organism is always translating any new experience through a transcontextual combination of the experiences that have come before. There may be a time of day you linked to the pain of heartbreak, a friendship that connects to a particular book.

These referencing combinations are beyond rational, as they hold our inner worlds of experience of diffused experience at nth-orders. Which is to say, there is no "clean" hypothesis. All hypotheses will carry in them something like a fingerprint of the organism's experiences that are hypothesizing. "It is not just that and nothing more." There are many contexts of experience that subtly blend as they form the basis of an idea to be expressed. The expression and the idea will be soaked in these preliminary processes, while the observer has no control or knowledge of this.

The unseen here, the aphanipoiesis, is working underground, mixing and matching differences, generating impressions and stirring them into the amalgam of history, anatomy, and physical context, and fusing those differences into other relationships the observer has experienced. The aphanipoiesis is taking place, producing implicit conceptions out of all the clutter and gems of past perceptions, most of which were not consciously construed. In this sense, there is what could be imagined as a pre-anticipatory field of necessarily non-prioritized experiences, waiting just in case.

An event or something communicated—whether a falling leaf or greeting of a friend—is not at all what is received as information by another being. There is no way to separate the observed from their history of observation, and it will always shape the communication to a form that it can digest. While this may appear to be a permanent confusion, it is possible that the aphanipoiesis, the unseen coalescence, is necessary for preserving the wildness of change.

COMBINING

Expression, Communication, Metacommunication

> When we study the biological world what we are doing is studying multiple events of communication. In this communicating about communication, we are particularly interested in describing injunctions or commands— messages that might be said to have a causal effect in the functioning of the biological world— and in the system of premises that underlies all and makes them coherent. (M.C. Bateson & G. Bateson, 1988, p. 151)

Mutual learning and information create resonances between entities in a living system. The resonances become their communication as well as their possibility for communication. This is perhaps at the core of what is sometimes called "change."

Explicit communications, such as when I tell my children I will walk the dog, or when the dog brings me the ball to throw, or when the news anchor says the vaccine is 70% effective against the coronavirus, all carry multiple versions of communication within them. As discussed above, there is a difference between what is "said" and what is "heard." But there is also another sort of communication implicit within each of the above examples that is non-trivial. When I tell my children I will walk the dog, I am speaking into multiple contexts. One context might be that I will do a task they should have already done; it could be a guilt trip. And it is also communication about our day, the activities it contains, and how we keep track of things. It could be a message about getting exercise; after being on screens all day, I might be modeling a healthy lifestyle. Maybe I am just eager to walk in the woods, which means the kids are watching the house while I am out, and they better have the dishes done when I get back. "I am going to walk the dog"—a simple statement of regular action, is, in fact, a loaded communication about the way our family communicates. Implicit in the statement is the message that "this is how we communicate about this."

Implicit communication is what I would like to address here. Firstly, the tone or the aesthetic of the communication is holding a great deal of the message. Aesthetics of relating in systems seem to become a grammar into which all other communication is soaked. The differences in tone and contrasting aesthetics of communication produce a felt spectrum that holds the limits of what can be communicated in any particular relationship. So if one were to read the transcript of a family interaction, it would be impossible to discern the nuance of the tonalities. Additionally, it would be difficult to discern those nuances if one were analyzing the family in person, as the observer's filters will not be congruent with those of the family, which holds history.

The meta-messages lurking in spoken communication, architecture, media, cultural expression, arts, technologies, and intergenerational expectations create the territory where relationships can be explored. The meta-messages of the classroom assign the authority of the teacher. The meta-messages in my Google calendar tell me that my productivity is seen through blocked-out timings. My mother's freshly set table communicates that I should sit up straight and engage in a more formal conversation at dinner. These are the unseen, unsaid, implied limits into which the relationships can expand. The aphanipoiesis is holding this meta in place, or maybe it is more insidious, depending on the family, or classroom, or forest. Are those meta limits protecting the possibility to learn and heal? Or are they delineating forms of holdback necessary in the system that devitalize? How can we know the difference?

Communication is not what has been said. It is what it is possible to say—and that is guarded by the implicit messages inherent in the relationships. Aphanipoiesis as a concept places this realm of potential communication as a territory of changeability, but it is not accessible from any map.

Warm Data Labs

Observations of aphanipoietic phenomena are surfacing through research conducted by Warm Data Labs and the online Warm Data process known as People Need People. Warm data processes are transcontextual mutual learning sessions open to anyone of any level of education to participate in. They are hosted around the world by hundreds of Certified Warm Data hosts who have undergone a study grounded in many Batesonian theories. The Warm Data process consists of a question offered to the group, who will discuss their thoughts in stories or other impressions as they move through multiple contexts. The question might be, "What is continuing?"—and the contexts might be ecology, education, economy, health, family, history, identity, technology, religion, and spirituality. As participants move through the contexts, their inputs begin to intertwine, fuse into new insights, and reframe memory. But it is not so easy to pinpoint where the change is taking place, or how, or to what end. In this form of conversation, people find that as they discuss their ideas "about" the given contexts, how the contexts link into their lives is revealing a form of learning happening "within" and "between" the contexts. The relationship between the "about" and "within" of the conversations in these practices has been fascinating and essential.

What has become increasingly clear through these processes after hosting hundreds of Warm Data labs with thousands of people is that it is necessary to re-examine what is meant by "change" in living systems, such that the change might be distributed throughout the system in unseen ways. Instead of isolating cause and effect, goal and strategy, to produce a particular change that is explicit and perhaps measurable, there appears to be a realm of potential change, a necessarily obscured zone of wild interaction of unseen, unsaid, unknown flexibility. The potency of this change is easily dismissed because it does not show up on the report with coherent analysis. This sort of change seems to eschew analysis. In fact, analysis as we know it is not suitable for studying this sort of slippery poly-learning. The learning is "actionized" differently for each participant; these sea-changes often move silently into many aspects of life, from professional to personal, without being traced back to the Warm Data work. The shifts in perception run deep enough that they are felt to have been there all along and are simply woken up by the Warm Data or, better still, continue without any mention or reference to Warm Data. They submerge.

The way in which Warm Data Labs both imitate life and disrupt cultural segregation of contextual attention is significant to this study. As participants move between contexts having conversations, they zoom in on the details of their particular context—education or ecology. But they also "know" they are in a transcontextual setting as economy, health, culture, and politics are also vaguely "there," producing conversations that implicitly achieve simultaneous zoom-in and zoom-out.

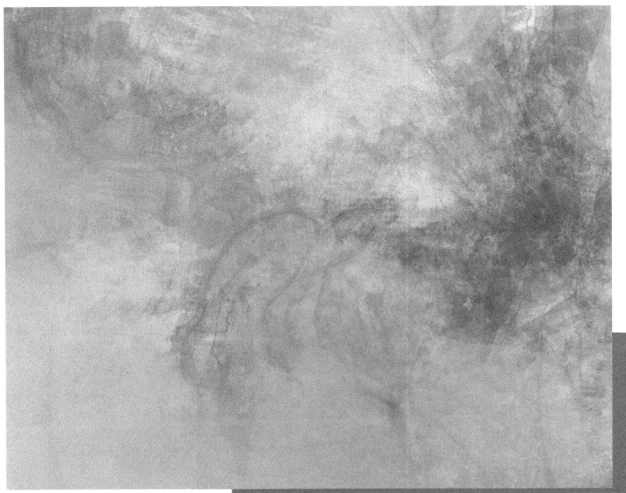

Fig. 28. Vivien Leung. (2022). *Untitled*. [digital art].

COMBINING

How the original question asked in the Warm Data Lab unfolds through each context is utterly personal and unique to the moment and the people in the conversation. The detail and intimacy of these conversations reconnect the broader systemic view into the participants' particular life experiences. This reconnection is critical.

The structures of systemic processes become tangible in people's memories of their lives. Not as a vocabulary or model—but as participants move to another "context" and continue—the conversation they just had does not fall out of their bodies. Rather, it is there, informing the following conversation, infusing the next context's discussion with the flavors and memories that were stirred in the last. In this way, transcontextual learning is a re-tissuing of understandings between people as they move through contexts, memories, language, and non-language.

Again, where is the change?

Those looking for a focused action plan toward a solution to a particular issue will be frustrated by the Warm Data processes. They will be unable to perceive the shifts that are taking place. The change is in the dark—not where we can see it. One of the most important things I have learned through this work seems terribly obvious: one cannot explicitly change what is implicit. This is where the Warm Data work has been so difficult to defend—in a world that seems to have forgotten the potency of the implicit. It is also juicy.

CHANGE IN IMPLICIT REALMS

As an entry into the vocabulary of systemic studies, aphanipoiesis is not so much filling a gap as honoring the need for gaps in all messaging to enliven the connectivity processes of communication, play, and learning in relational process. The gaps are essential; they hold unseen, untamed, unnamed, un-grooved mixing of collected impressions. Gapping is inevitable, and it is vital. The relational organism cannot help but engage in casting perception across these gaps to find the familiar in the unknown. With no goal other than to continue to be a living system within other living systems, the organisms are free to mutually follow stochastic pathways as they happen.

> Now that we have pulled out "structure" from the ongoing organized flux of the universe, it is appropriate to attempt a synthesis—to put it back again. Let us see how our fabric of descriptions and reports, and injunctions fits a world fleshed out with life and happenings.
>
> First, it is conspicuously full of holes. If we try to cover life with our descriptions of it—or if we try to think of the totality of an organism as somehow fully covered by its own message systems—we at once see that more description is needed. But, however much the structure is added, however minutely detailed our specifications, there are always gaps. (M.C. Bateson & G. Bateson, 1988, p. 162)

Re-tissue-ing the Gaps

It is as though the gaps provide the necessary opportunity for tissuing, connecting, and impression-ing processes to take place. The stitchery between these gaps is the abductive process at work. In the same way that metaphors generate responses from the unseen inclinations of the observer, the gaps are there to allow themselves to be filled with inter-steeping inklings. In that stitchery is where the rhythm, the tone, and the rules of communication between aspects of the system are forged. Later, those rhythms and rules manifest as emergent behaviors or events. By then, the implicit underworld that defines the relations has taken form. The question that aphanipoiesis brings is how to tend to that permeable area of stitching the gaps.

> In sum, all descriptions, all information, is such as to touch upon only a few points in the matter to be described. The rest is left uncovered—hinted at perhaps by extrapolation from what is actually communicated but, in principle, undetermined and uncontrolled by the message system. The US Constitution, for instance, leaves almost everything unsaid. What lawyers have spun out, in addition, still defines only a few details and, here and there a basic principle of human interaction. Most is left undefined or is left to be worked out after the first formative hint is given. (M.C. Bateson & G. Bateson, 1988, p. 163)

What happens between hints? Between which tendencies and habits were the connections brewed? Explicit actions fail to touch this process. You cannot tell someone not to be racist, or corrupt, or sexist, or greedy. Those insidious ways of being in the world are produced through a conjoining of experiences and unseen impressions. They are out of the reach of any direct corrective.

Aphanipoiesis must be taken seriously and handled with the utmost integrity to get to these implicit realms. Art, play, practice, learning—all of these are natural responses to the need to try out new ways of weaving connective ideas and reweaving them and reweaving them. Placing images, musical notes, ideas, stories, and other forms of open communication side-by-side is an irresistible invitation to start finding new links. The tissues between the gaps hold the implicit premises of the system; they form the most malleable and perhaps the most challenging realm of systemic change.

How does one know what information to put "side-by-side" to generate stochastic stitchery that is aphanipoietic rather than insidious? It is a razor's edge of "difference that makes a difference." Stochastic process vitalizes the unseen stitching between gaps and utterly depends on abductive process.

Precious unexpected insights that bubble up between contexts offer a side door out of the matrix of self-perpetuating thinking loops . . . but it is not as methodological as it sounds. In the Warm Data work, there is an uncommon attention paid to the levels of abstraction the contexts share that are being placed "side-by-side." It matters. Bertrand Russell (1908) and Alfred North Whitehead (Whitehead & Russell, 1997) both worked with the idea of logical typing; Korzybski (1994) entered this discussion as well in his inquiry around levels of abstraction. Mixing the levels is a dangerous game.

As the aphanipoietic process is always taking place, there will be, for better and for worse, a stitching together, a coalescing between the gaps, whatever they may be. The question is—how much transcontextual tissuing is possible? If the gaps provided are limited, the coalescing will be impoverished. This is the basic tenet of what might be seen as insidious propaganda. Abuse of the aphanipoietic process would include narrowing the contexts through which abductive process might be able to bounce around, coalescing increasingly complex sets of gaps/contexts. The limitation creates a basis of a reduced living process and leads to more reductionism, divisiveness, and violence.

In general, as we are becoming sensitized to the nuance of how the abductive process is activated through this "side-by-side-ing," it is best to keep the abstraction levels congruent. By that, I mean that the combinations are lateral. We might side-by-side contexts such as politics, economy, health, family—but to add something like "love," "communication," or "competition" would bungle the levels and likely create confusion.

For example, comparing nature to a machine makes it possible to see in nature a fragmented collaboration toward an outcome. This, in turn, begins to justify competition, war, and individualism. Whereas if the side-by-side-ing places family, ecology, and economy next to each other, what starts to happen is that the deeper planes of story reframe into an understanding of what is non-trivial, profound, and vital.

The "what" of what is being placed side-by-side by a Warm Data practitioner hoping to offer insight shifts to implicit realms is critical and should be addressed more fully in another study. I must stress how delicate this is; the seduction to place an agenda or predefined outcome to this side-by-side process is but a return to the manipulations of propaganda, which is why my father regretted the work he did with OSS during WWII. (More on this in the *Ecology of Communication* chapter towards the end of the book) The lure of urgent transformation as a controlled process is inherently abhorrent to the wildness necessary for aphanipoietic change.

Re-wilding the Interior

The layout of side-by-side contextual process is where we tend the explicit to make room for the implicit to shift. In the words of the Warm Data hosting theory, we tend the "about" so that what is re-configured is in the "within." It does not really matter what people talk "about" in a Warm Data Lab. There is nothing to capture at that level. What matters is the way the participants are internally sewing together the different conversations and contexts. On a transcript, this information is inaccessible.

In the Warm Data processes, communication in explicit form is not held to be the communication of interest. That level of conversation is there as a skeleton, onto which the stories not told reshape the person who did not tell them, the alterations in tone, the re-tilted perception is given free rein to rub memories and stories against each other. One comment that comes up repeatedly is, "Your story changed my story." Through this "side-by-side-ing," stories told change stories almost told, and their bearers are able to reshape their impressions in ways that are untamed. By careful tending of the "about" and "within," the rich world of memory and story re-wilds.

The gaps are where the hope of systemic transformation is waiting. In the Warm Data processes, participants are given a structure to re-stitch, to re-wild, and to begin a new abductive process into these gaps. Again, by placing the contexts of life side-by-side in new configurations, the aphanipoietic processes are given room, without conscious purpose or goals or defined outcomes, without scripts or roles or trends—to allow the tender new beginnings of another abductive description to form mutually. Through this work, I have found I needed this term to embark on a deeper study of the importance of aphanipoiesis. The changes I witness occurring in the Warm Data processes are completely unpredictable and profound. They suggest ever more vividly that there is a real, if unseen, mingling of the body, culture, education, family—and a whole batch of transcontextual experience guiding all other actions. It is to this change that I have devoted my efforts toward systemic transformation.

What Are the Implications of Aphanipoiesis?

Aphanipoiesis offers an introduction to, and an invitation to further explore and develop, a theoretical basis from which to address all that coalesces prior to emergence. The word is a description of the process, but it is not the process itself. The name is not the thing. This theory is not meant to imply a way to find and expose the unseen but to approach change differently. The unseen coalescence is also ungraspable—in a culture where change is entangled into an eagerness to grasp, define, analyze, and take action. When change is sought through adaptation to existing systems, it is sourced from the system itself. In this case, perpetuation is more likely than change. However, when the existing structures are not present, organisms must "find a way" by sourcing from unhabituated living, shifting impressions that have been brewing over time. The likelihood is that this existing structure will continue, grooved into habituated responses that often obscure those that have not been practiced.

Aphanipoiesis introduces a new paradigm and theoretical approach to change that sends us to the forest to find what we have lost, even if it is dark. As a concept, aphanipoiesis offers permission to take seriously the significant change taking place in ways that are unreachable by analysis and direct action. This is a moment in which there is a dire need to become and live into ways of being that cannot be informed by the existing structures.

How do human beings or other organisms know how to change into something they do not know how to become? This flexibility for transformation is waiting outside the familiar. The task now is to create the conditions for vitality—the conditions that encourage what might be recognized as re-wilding in biodiversity in nature—but in humans, it could be tending an ecology of ideas. Let them be many and filled with movement that allows new contact—re-wild the aphanipoietic realm.

Ready-ing—Tending the Prelude of Change

When an organism responds to an event in its environment, it responds from the combining cauldron of experiences that have formed and in-formed it. Yet, to continue "to be" requires constant shifting and learning to be in new ways. In our exploration of "systemic change," it would seem that there is a process prior to the "change" that allows organisms to become ready to respond in new ways. This process of ready-ing may be what makes the subsequent change possible. Before the change, there is a prelude, a priming, a saturation of mutual learning between organisms through which pathways of possibility are produced; we can imagine this as a process of becoming ready to perceive and thereby respond in new ways. What does it mean to become ready for change? This process is open-ended, always sensitizing, and ever-learning and takes place within an already existing aggregate of perceptions. The limitations of what is possible to become ready for are found within the potential combinations of previous experience and contexts with the new experience. Into what teapot of complexity does the new arrive? To be ready is not a definable state; after all, the unripe mango is the one ready for particular sour flavors to occur, whereas the sweet, juicy mango is ready for other flavors.

Readiness is also not linear; it is produced in unpredictable transcontextual ways; learning to carve wood can bring readiness to paint or fix machines; a broken heart can make it possible to understand a story or song in a new way. Attention to this process is rigorous; at the same time, it is ecological in its scope. Ready-ing is slow and sometimes fast; soft and also harsh at times; it is seen and unseen. To give it room to move is a kind of humble preparing toward undefined multiple pathways. As participants in this process, what might it mean to tend the possibilities of ready-ing so that whatever actions unfold do so less destructively?

The opposite of ready-ing is to force a single-purpose outcome upon a system oriented and shaped around pathways that do not correspond to the desired change. Natural processes are never singular. Ever. To dictate an action change upon a living system that is not "ready" brings contortions of false and unrooted change, which falter, splinter, and become grotesque. An unready change is a violence in so much as it may contort the process of learning into place before the necessary undergrowth has taken place. This may lead to insidious difficulties later on.

For instance, one cannot make a rule that people "must respect each other" and expect that the respect will go beyond vocabulary. Respect cannot be mandated; it is produced in the specifics of context, in unique relationships, and in the un-modelable complexity of each person's experience. False respect may be put into visible practice as rules to abide by, but mutual disdain may still infect the relationships in unspoken ways. For example, telling someone to stop being depressed, to love another person, or to be trustful is likely to produce odd configurations of the suggested changes.

The unready change is a short-circuit of the necessary transcontextual mutual learning in ways that often lead to deeper insidious difficulties later on. The "goal" of a linear strategy may be to generate respect, optimism, or trust—but the culture of strategic productivity is ill-equipped to meet the wildness of life. To produce the goal without the ready-ing undermines the depth at which respect, trust, and optimism are the consequences of relational possibility, not mental states to adopt.

Suppose one sees the world through the lens of strategy. In that case, it will be challenging to produce the conditions for respect, trust, or optimism to occur because these characteristics must be natural in their arrival: untamed, unscripted, and allowed to shape as is possible in the particularities between people. In a shared culture addicted to "endpoints" and "outcomes," the communication of that untamed possibility is incoherent.

Like all ecologies—which are vitalized by their interbeing—the ecology of ideas shared is constantly forming interdependencies of their own. This includes cultures. An ecology of ideas can be as alive as a meadow or as destructive as a mafia. Either way, the organisms depend upon the continuation of the relationships. The circumstances may change; there could be a drought that disturbs the patterns of the forest or a pandemic that disturbs the habits of society—the relations within the ecologies will seek new ways to stay the same.

Getting out of a Möbius strip of epistemological looping is a sticky problem, especially since the concepts of how to get out are produced from within the loop. This is where the depth of the industrial, engineering-based notions of explicitly defining and strategizing change in terms of seeable "results" or "progress" is perpetuating the thinking that is producing the transcontextual crises that a continuing human species will have to get out of. In industrial terms, the product became a linear endpoint towards which a culture of how to get to that desired endpoint or output was placed upon a race track of faster-more-cheaper. This has become wrapped into education, economy, health, and technology in ways that have soaked into identity and language. It will not be easy to effect a shift in this thinking, though a more ecological version is necessary, one that is not fast or premised upon more or cheaper but instead attends life.

Another approach to the idea of change is needed. Below, the preliminary gestures of such a shift in approach are suggested—leaving the familiar world of mapping change, measuring change, defining intentional change, and even specifying change—instead, we must tend to the unseen flexibility of living systems that produce a readiness for change. We must move from change as an outcome of particular actions to change as a consequence of systemic ready-ing.

The difficulty will be that there is no way to know what change has occurred or in which direction. When working with complex systems, it is imperative that this movement in 2nd-order and nth-order change be an accepted understanding of how change takes place. Pre-scripted outcomes cloak the ability to perceive transcontextual change. This is not a comfortable position for the current epistemological understanding of change, which requires predefined outputs.

Knowing How to Evolve

Let us begin with the question, "How does an organism know how to evolve?" This question, posed by my father, is a deconstruction of a collection of presuppositions about change. The first crack this question reveals is the tautology of the familiar: An organism does what it does because it does what it does. The bees are bee-ing. The earthworms are earthworm-ing. How would they know to do otherwise? In fact, the ecological interdependencies these organisms are interlocked with also rely upon them continuing their way of living. This would indicate that continuation would require that every organism keep doing what it was doing—and yet evolution is survival. Evolution requires that all organisms contain and nourish flexibility for constant change. To continue to be a meadow, every organism in the meadow must contain the possibility for unfamiliar shiftings in relationship with each other. Actually, these relationships are essential for change processes. What a fantastic headache.

Combining

William Bateson, Gregory's father, said in a letter to his sister, Anna, in 1888:

> My brain boils with evolution. It is becoming a perfect nightmare. I believe now that it is an axiomatic truth that no variation, however small, can occur in any part without other variations occurring in correlation to it in all other parts; or rather that no system in which variation of one part had occurred without such correlated variations in all other parts, could continue to be a system. (W. Bateson, 1888)

The question of "how does an organism know how to evolve" leads to inquiry of another realm of communication and information beyond the habits organisms are sensitized to and into a gooey realm of combining experiences that must be outside the familiar. If this realm is sensed within familiar sensings, it will be metabolized into existing ways of perceiving and responding contextually. The new limb, the new idea, and the new ways of being are bubbling between and under and to the side and out of reach of the organisms' habituated perceptions.

> It is out of the random that organisms collect new mutations, and it is there that stochastic learning gathers its solutions. Evolution leads to climax: ecological saturation of all the possibilities of differentiation. Learning leads to the overpacked mind. By return to the unlearned, and mass-produced egg, the on-going species, again and again, clears its memory banks to be ready for the new. (G. Bateson, 2002, p. 45)

Yet, in order to enter into substantive inquiry about this kind of systemic change, it is critical to address the limitations within research methodologies that cannot account for nth-order change. As noted above, William Bateson found that the responses to responses to responses of multiple and simultaneous contexts are obscured when reductionist outcomes are required. This has been a problem for systemic studies and a friction that is perpetuating measurement and methodologies that can only track 1st-order changes.

Behavior change in systemic process is especially difficult to articulate in terms of causality and directionality. The importance, therefore, of addressing the nature of systemic change is relevant to the inquiry around aphanipoiesis, which, as a theoretical basis, offers the possibility of discussion and further exploration of unseen combining responses that generate systemic change. It remains difficult to predict or even qualify the direction of this change.

The issue is that the movement within the system is neither quantifiable nor qualifiable; it is beyond the reach of these two tropes and is better described as a transcontextual coalescence. The rigor of this description is not to be underestimated as it demands not only a multi-perspective approach but also a multi-perspective approach focused on the combining, not just the parts. The combining is alive, unstill, and reshaping through many pushes and pulls in the system, most of which are unseen.

WHERE DOES THE NEW COME FROM?

Change or learning in an ecological aggregate of relationships is a process of continuing while transforming in many contexts. Gregory Bateson (2002) writes:

> It thus comes about that what I have called double description becomes double requirement or double specification. Possibilities of change are twice fractionated. If the creature is to endure, change must always occur in ways that are doubly defined. Broadly, the internal requirements of the body will be conservative. Survival of the body requires that not-too-great disruption shall occur. In contrast, the changing environment may require change in the organism and a sacrifice of conservatism. (pp. 134–135)

Double description or multiple description is abductive, allowing life forms—and ideas—to connect into multiple forms. An earthworm is in relationship to the soil, the bacteria, the trees, the birds, the plant life, the insects. Each relationship takes form in very different ways, but the inter-stitchery of its form is what keeps the creature viable in an ever-shifting ecology. It can be argued that without an understanding of abductive process, there is little chance of any change that goes beyond perpetuating existing habits in the system. Abductive process is the fundamental theory of where and how the re-coalescing of the experiences through which pathways of possibility are perceived takes place.

> What seems to be the case is that there are, in nature and correspondingly reflected in our processes of thought, great regions within which abductive systems obtain. For example, the anatomy and physiology of the body can be considered as one vast abductive system with its own coherence within itself at any given time. Similarly, the environment within which the creature lives is another such internally coherent abductive system, although this system is not immediately coherent with that of the organism. (G. Bateson, 2002, p. 134)

In our study of ready-ing, the notion of what has informed an organism's making of a hypothesis is central to what perceptions will be possible to let in their way of seeing their world. When considering where readiness is formed and informed, abductive process is the coalescing zone, the cauldron, the meeting place of impressions that reshape each other. The experiences and formations of possible perception shape what is perceived, making the hypothesis that is produced an indicator of what came before.

Peirce's notion of the hypothesis is therefore joined with Bateson's criteria of mind to reveal the particular brew of experiences into which a new perception must attach. Aphanipoiesis addresses this through the idea of a constant coalescing of perceptions and experiences that are not distinguished from one another but rather form a hum or a resonance of meaning-making into which a new sensation or idea lands.

Combining

Fig. 29. Vivien Leung. (2023). *Searching*. [digital art].

Fig. 30. Vivien Leung. (2023). *Life Spiraling Art.* [digital art].

Fig. 31. Vivien Leung. (2023). *Mama Blobs*. [digital art].

IT'S A GAP

It's a gap,
That is holding our fibering.

It's a gap,
That is both eachness and reaching.

It's a gap,
That is our learning to find one another.

It's a gap,
That pulls us and pushes us.

It is - - - - - - gapping.

Please do not complete me.

Let me relish longing.

More on Side-by-Side-ing: The Necessary Gaps

Abductive process is made possible by gaps. The organism, by virtue of being alive, makes connections across the bridges of experience. There is no way to stop this process of stitching the gaps together into an epistemology or an umwelt. This is what living things do; they perceive relationally. The gaps are the place where pure possibility sits. It is in the connective process that the tautologies of the familiar have opportunities to rearrange. In our previous question of "How does an organism know how to evolve?" the potential for doing and being what it has no previous experience of doing or residing in the gaps—the re-stitching of the gaps—the abductive process through which new descriptions and hypotheses happen. The ways in which an organism learns to be in its world are overlapping throughout the relational interdependencies it lives within.

In today's human terms, this looks like the way a preschool-aged child learns to write their name on the paper, when to snack, where to play, who has authority, and so on. These relational shapes are repeated in many other contexts of the child's life and into adulthood. The same learnings are there in the workplace, the doctor's office, the police station, the sports teams, and the rules of the games. The overlapping and transcontextual repetition of these rhythms of relation forms a "grain" in the wood of "what is real" for the person.

How are we learning to be in our world? This question points not only to the individual organism as above but also toward the ways mutual learning holds systems together. The collusion of shared learning becomes a communal set of possibilities into which the organisms are acting. Ecologies are only possible when multiple organisms are continuing in interdependent communications. To do so, they are in a continual movement of ongoing learning through that communication, always shifting the limitations of what the relations can be. Stasis is obsolescence, whereas homeostasis is continuing to learn. In order to continue to stay the same in some form, organisms must change in others.

Again, the tree, the bacteria in the soil, the earthworm, the birds, the meadow flora, and the insects are all mutually learning to "meadow." By the same mutual learning processes, children, parents, teachers, employers, workers, grandparents, lovers, and friends are all mutually learning to continue life as we know it. We identify ourselves and each other through the signals of job title, material wealth, physical fitness, and status cues in communication and gesture—and we respond to those collective agreements either by complying with them or rebelling—but either way, we are still in relation to the knitted knot of mutual learning that "we have learned to be in our world" within. So, how, if organisms are holding each other in the tautologies or matrix of their knowing—how does change happen?

Perhaps one of the more profound responses after a Warm Data Lab is the observation, "Your story changed my story." The actual change is not articulated, but the observation is. This is what it looks like when a change is happening in an abductive process. The specifics of the change are not graspable, but the general shift is widespread throughout the system. This is significant because it reveals that when a personal story is told, the listeners receive it through and into their own collection of experiences. The listeners have impressions and memories, form responses and un-form them, and make connections to their own lives in ways they might not have if another story had been told. The impressions are salient, not to be spoken of or harvested, but to let them flush through the habituated grooves of familiar exchange. Then, when the listeners move to another set of contexts and other participants, more stories are told, releasing more impressions, memories, and unformed thoughts. These amorphous bits and flashes begin to find each other, shape each other, resonate, and reply to each other in ways that are entirely unpredicted and unpredictable. In the aphanipoietic realm, this combining happens through inputs side-by-side that reknit, coalesce, and blend. In the aphanipoietic realm, this is a process that happens through having inputs side-by-side that reknit, coalesce, and combine.

Here, the tautologies are tickled, and the habituated collusions are mutually learning to be in new shapes. The matrix gets remade. The "new" comes from the gapping—the side-by-side-ing—the way an arrangement of images or experiences, senses, or inputs produce relevance between them. Like a conversation in which the meaning is not said but is made through the spaces between. For those looking for takeaways or harvested insights, this can be frustrating, but the real shift takes place after the session, not during when the participants find they are responding in new ways to personal, professional, or even physical contexts.

I'm thinking about a woman in a Warm Data Lab recently who said that afterward, she was sitting in her living room. She looked around, and she noticed a lot of books in her library that were not hers. Then she said, "What kind of a person does not give books back that they borrow?" For some reason, she was suddenly able to ask herself this question. After all, some of these books she'd had for years by then. She was now able to think about those books and the relationships they came from. Then she started giving them back.

And—when she started giving them back, all sorts of other things in her life started to change. She started to perceive beyond the books into all sorts of other contexts. Something was changing in how she was in relationship with her world and tending to her integrity with others—if you want to call it that. Or, if you want to just think about the tenderness or the mutuality, whatever words you want to use—there was a shift. Immediately, she noticed the books as having opened this pathway for her. Now, I can promise you that in the Warm Data Lab, there was no discussion of people returning books, nor was there any discussion of what makes you a good person or a bad person, nor was there any indication that this process was there to improve your moral compass somehow. However, the Warm Data Lab process of connecting different contexts in her own life made it possible for her to see those books in a very different way, which she did not expect, and nobody else did either. On the one hand, this book story may seem insignificant to a world in a polycrisis. We do not have a global crisis of people not returning books. But the survival question is, how do we live together in a new way? And that's where these shifts in the minutiae of people's lives are actually indicative of much deeper shifts, I think, are very profound.

CHANGING CHANGE

A theory of change can be an invitation to action with the intent to produce linear, mono-causal, predetermined outcomes or goals that can be both labeled and measured. This linearity is an epistemological habit of the industrial world. The irony is that the most pernicious problem facing humanity now is the habit of attempting to respond to living, non-linear, multi-causal issues with responses geared toward specific outcomes. To do so is to produce actions fundamentally out of sync with the livingness of the problems. There may be truth to the old saying, "You cannot solve our problems with the same thinking that created them." Generally, the change in a theory of change is thought to result from providing inputs and transforming them through action, thus yielding outputs and outcomes. But—what if the change precedes the action? What if the more significant aspect of systemic change were, in fact, a prelude to the inputs and actions, something more like a readiness through which an unforeseen action can become?

Fed up with promises of change-making that seem only to perpetuate the existing systemic issues, exploration of the theory of aphanipoiesis has begun as a basis of another order of change. The early inquiry into this has been a fascinating and rewarding discovery process of the nuances of systemic change, as well as the epistemological traps of presuppositions about change itself. Especially in times of urgency like these, the habit of envisioning a solution to a "problem" pulls most of us into a linear imagination. It is difficult to hold back the urge to strategize, create action plans, and project a preferred outcome as a goal to work toward. This, of course, is the exact thinking of pre-planned action to solve a problem that got us into the mess in the first place.

Is it possible to imagine the necessary change? This is a non-trivial question—pointing to the predicament of epistemological habits repeating themselves. Imagination is not formed in nothing; it is brewed in the experiences of the organism. From that pool of experiences and relational limitations, a hypothesis can be made as to what change is required. But the imagined goal or direction of change is dripping with existing presuppositions; some are obvious, and many are not.

Because the abductive process that is the brewing place of perception of change is un-seeable, aphanipoiesis provides a way to understand the unseen coalescence as a necessary aspect of how living systems continue in multiple interdependent relationships simultaneously. In this sense, the word "learning" and the word "change" are held to be equivalent. What has become evident is that most theories of change seem only to perpetuate existing systems of thought and action. In the face of the current urgency, it may be necessary to explore outside of the realm of the familiar and into a realm where flexibility for potential change is produced.

CHANGE IS STUCK

The pain and horrors of exploitation seem to be outside the reach of activism, policy change, laws, and moral mandates of religious practices. Like addiction, the continuing practices of the existing systems are shaped in the premises of insidious ideas that are so deeply woven into the basics of daily life as to be waterproof to demands for change. Premises that generate their own logic of action, such as individualism, productivity, wealth, success, efficiency, progress . . . If these are dyed into the basis of a collective understanding of how to live life, they become the premises, and from there, no eco-law or corruption law will alter the course of the group. Instead, notions of ecological change, societal change, or cultural change will be inducted into the "individualism," or "efficiency," or "progress."

We see this in the current irony of "collaboration" as a buzzword of change-making to the extent that there are even competitions to which people can submit their ideas (e.g., in 2017, the Global Challenges Foundation in Sweden offered US$5 million for the winner of the "Collaboration" contest). Of course, those people cannot collaborate on their submissions and may find their ideas have been poached by the other "collaborators" or so-called green businesses caught in the cycle of planned obsolescence to keep up sales. Or the popular notion that increasing energy efficiency—or, more broadly, resource efficiency in business—is a desirable way to reduce humanity's environmental footprint, only to discover that greater efficiency ultimately reduces costs and product prices, which in turn stimulates demand for the company's products with uncertain effects on the environmental footprint. Or company by-laws that demand equal rights for employees, transparent investments, and ownership but whose product is produced with child labor and is financing terrorist organizations. The list is endless. The system is premised on a scaffolding of ideas that reproduce under the auspices of change but, in fact, perpetuate the same problems it seeks to change.

It is the whirlpool of spinning epistemological repeat—a tautological bondage; it is the knowing that prefers what is known. It is the trap of the familiar. Somehow, still, life keeps moving.

X + 1: Time Is A Variable

How long does ready-ing take? How does one know that anything has happened?

What are the indications or markings of ready-ing? There are none. And there should be none because this process takes place through and between many aspects of mutual learning, memory, communication, and unformed, unseen coalescing; no one will ever know where, how, why, or even in which direction ready-ing happens. It must remain wild. The alternative is a push or manipulation of becoming ready that will be formed in forced ways. Where the ready-ing is not wild, there will be an underestimation of the multiple processes needed to allow the stitching and combining to bring forth.

Human beings are often condemned to the perception of their world, which is dangerously lacking the millions upon millions of other organisms changing each other's stories. In that reductionism, the perceived "need for change" or the perceived ready-ing will likely beget something like Frankenstein's monster and not a viable living learning process. When the determined "purpose" is not complex enough, it will twist and contort into un-ready ready-ing.

One cannot simply say "trust is needed" and hope that trust will be formed. The relationships are not changed at that level, nor are they changed at the level of demanding new communication. Rather, they open to the possibility of 2nd-order trust only when the possibility of communication itself is changed. This takes an unknown amount of time, an unknown combination of impressions side-by-side-ing, and an unknown mixing of perspectives. This is referred to in the Warm Data work as the x + 1 factor: x amount of time plus one minute. You never know where you are in x, but when the one minute happens—something shifts everywhere.

Combining

The movement from a theory of change that is located in "action" to a theory of change that is located in readiness is a shift of attention from what can be "done" to achieve an outcome to how to nourish the conditions into which the organism learns to be in its world. The readiness is not findable or definable by anyone. Most of the time, in the case of human beings, the person has no idea whether they are ready or what they are ready for.

Some readiness is like a ripeness, whereas another ready-ing is not. Sometimes what is needed is unready—the voice of the children, a sapling—something about this is "sacred" territory. It is truly alive and alive with what makes life keep life-ing. Attempting to manipulate, predetermine, or shape this aliveness is a violation of its own possible learning. Rather, we can create the conditions for transcontextual mutual learning into which new abductive processes are possible. From there, life makes itself.

Conclusion

Most of the limits binding the possibilities of change are ingrained in the unseen premises of how life is shared among organisms, human or otherwise. The interdependencies are in an ongoing continuance as is necessary to their ecology. The unseen aspect of possible movement and change offers a parallel realm from which possibility can be drawn in times of change or stress. This aphanipoietic realm is where the ready-ing is waiting, forming, and informing. While it may seem as though the objectives of change should be clearly defined and mapped out, or that ready-ing can be achieved with direct purpose in mind, the transcontextual movement and learning within and among organisms seeps into other directions of relationship. This is a good thing. Irritatingly out of reach perhaps, but still, this realm is full of yet unseen and undreamed of possibilities. The difficulty is in the urge to immediately instrumentalize and harvest this possibility. Notice that this urge is rooted in the premises of the same actions that are generating the predicaments to begin with. Ecologies perpetuate themselves, and ecologies change. Aphanipoietic movement is something of an underworld of other ways of life life-ing that waits, unseen. This underworld is full of teeming, swirling, incoherent futures. Let them brew; they are hope itself.

Acknowledgments:
I would like to thank Phillip Guddemi, Leslie Thulin, Tim Gasperak, Andrew Carey, and Lance Strate for their care and attention in reviewing this document in its content and form. It took many sets of eyes and many minds to produce this paper. The learning described in this document is a culmination of decades of study, generations of experience and, most recently, it has in part been made possible by the whole Warm Data group and the thousands of Warm Data participants.

---

* Previously published in *Journal of the International Society for the Systems Sciences*.
  "Aphanipoiesis" was changed to include a section of the paper "An Essay on Ready-ing: Tending the Prelude to Change."
  "An Essay on "Ready-ing: Tending the Prelude to Change" was previously published in *Systems Research and Behavioral Science*.

LISTENING TO THE LISTENERS

Time to be listening to the listeners ...

The water is certain of her stones, their ecstasy, and the tingle of re-greeting with the tide.

Stone flesh hiding time in mornings and evenings.

Ancient.
Softening.

Rolling over and over in long measures of colors over-folding into a chord.

Reverberating through now.

NOTICING

The basic premises of how each day is lived into existing systems is in relationship to time.

When and what a meal is,
With whom the food is eaten,
The pace of conversation,
Breath,
Heartbeats,
The laundry schedule, the shopping hours, holidays, walking the dog.
The tempo of family, shoes, and coats, in the doorway, coming and going, the work day commute,
The compost bucket is full again.
The bills need to be paid again.

Habits stitch the rituals and smooth assumptions of what is needed materially to produce the next day,
Moment-to-moment clocking and banking of "time spent."

What happens when the relationship to time changes?
This is the secret undergrounding of unseen changes.

Changes whispering to other changes in Morse code. The tempo shifts.

All of life is rhythmic.
Unfold contracted hours packed like foam mattresses in small boxes into the open air.

Soon they will never fit back into their packaging.

Change the rhythm – change the dance.
The chickens and the garden are still in ancient time.

KINKY

When you see the clarity of a future horizon –
Turn quietly into the thick bush.
The more elegant response is the one that wiggles.
Slipping from grasping anxiety, avoiding clean edges.

Time brings a way through the impossible.
But it oozes, slimy,
Entranced by the twisty,
sticky,
unwieldy bits.
The tangents, detours,
The curly pockets of crud and life.

The clear path is itself a warning –
Trimmed and tucked by Procrustean impulses of industrial habit.
Instead,
Find the vital tangle of broken lines and crags –
A fest of possibility in the festering
Societies of ideas –
decomposing.

Stinky belly buttons have more to offer the scouts now than a
thousand articles of strategic analysis.
Weird dreams, untidied
Sing the airy maps
So they will not be found ...
By the ones looking for management,
And numbers will mock their lovers.

Memories are rioting against reason.
The future won't fit into the fear of rotting.
It is the green fur itself.
The future is kinkier than we thought.

Fig. 32. (pp. 186-201). Vivien Leung (design and handwriting), Nora Bateson (poetry text & concept), Mats Qvarfordt & Trevor Brubeck @Handtryckta Tapeter Långholmen (wallpaper background), Rachel Hentsch (graphics), Leslie Thulin (graphics). (2023). *The Meadow Verse*. [digital art on images of handprinted wallpaper].

I want to be the soil that is alive with a world of organisms.

And I am.

I want to be the water that is the current of time

And I am

I want to be the green leaves tipped toward light

And I am

I want to be the rotting
log with fungi and insects
throughout
    And I am

I want to be the deep roots,
finding darkness slowly
in hard earth
    And I am

as we
all are.

Forests.

The day began with such unusual light. The sun was in a moon-ish mood. Like a form fitting dress, the atmospheric luminosity set off the shape of the sun's belly.

Again and ancient.

The miracle of the seductive tango between earth and sun.

Fire and dirt.
Light and water.

A wistful reminder of illusive warmth on a frozen December morning.

Creekside through water

From and
The slippery stones.
The leaf that is carried.

My blood
Your tissue
The roots reaching

Light

Paths to moon horizon

Still, trickle, gush

around obstacles.

Swifting

Chilled air with
earth tones

Turning and

Life itself

We are liquid.

The tall grass is still green
Summer is still nostalgic,
Even in these times —

The warm dust is weaving
yesterday's rain —
Like a tourist —
I romance the notion of
the wilderness.

Capture it in pictures,
as if I would otherwise
forget.

But I live here. Have I already forgotten that?

Stitching the soil back into my shadow.

Noticing there are pebbles in my shoes.

Trillions of organisms in and on my body are communing...

I am not a visitor, but always a host.

And — Always a guest.

Honor, celebrate, dance, love, sing, tend, embrace, write our fantastic membership, contemplate, draw, in the ongoing movement of living interdependencies of millions and millions of organisms learning and shifting as time swirls through us. Membership in ecologies of being.

In a blink

The whole forest

Is readying

To descend to the roots. And go underground.

I am lostness.

I am a soggy leaf on the forest floor,

I am a flapping billowed sail of a ship in a storm,

I am arms of pillowed warmth,

I am an earthworm.

I am sticky honey with lime and cayenne.

I am an old book with delicate pages.

I am nothing.

I am rooms of muddy boots.

I am in.

Light on water and stone just is-ing.

A leaf floats.

The year behind, the year ahead.

Small and epic stories pass. There is so much to do now-- there just is.

shake ancient geology

Pooled reflections cannot

Together they just are.

Things change. They just do.

Forest walking.

Throwing a big stick – for the joyous beast.

Unbounded bouncing with no because!

Just oh yes love, life play, go again,

It's not a digital thing.

But here we are.

CREATURE

The creature in me is screeching.

I am clawing through the metallic script delivered through the phone. I am gnawing through the numbered slots in the form. My organism is chewing through the wires and sabotaging the engine.

This ... in defense of the many versions of silence that I can share with another person, in which entire conversations can take place wordlessly.

This ... in defense of the deer that can sense a fox from across the forest after thousands of years of evolution, but not a car.

At my peril, I am stomping through the floor of institutional normalcy—like a toddler having a tantrum. We must reclaim our creatures to meet these delusional times of artificial intelligence and robots. While I can make economic and intellectual justifications for the cyborg sneaking into my skin, my backbone knows something is going horribly wrong. My creature is responding to the dread of a Kafka-like future of authoritarian technology from another response system. My blood and bruising flesh know this is something to scream about, even as I write this on my laptop.

What does it mean to take a stand in an interdependent system? To advocate for life, when to do so is to polarize day-to-day technologies and policies? How is it possible to consider the interdependency of the ecology when I have to relegate my complexity into violently reductionist contortions just to pay my phone bill or get a doctor's appointment? Suppose I am learning to embody the contexts of intertwining vitality necessary to continue life in the garden, the forest, the sea, my belly, and the music the woman upstairs is playing—how could I be a number? I have fingernails, shiny teeth, and tiny hairs on my arms. I can hide, blush, and attack. I can pee on my territory, even if I have forgotten for a couple of generations.

Speaking of change, how can I sensitize to new frequencies, as is needed to completely alter the social systems—while I sit here, on hold, choosing between six service options, all unrelated to my needs? The person on the other side of the line is unable to offer assistance. My request spans three departments, and she is bound to one. "Please send your request through our website or via email."

The clockwork of these structures has formed a sort of nasty nest of tautologies. The logic of the departments, the forms, and the limited selections prevail across schools, banking, health services, government, insurance, travel, retail, and even art galleries and theaters. This grid is spreading across our lives, promising order but delivering degradation of the human capacity to interact, to imagine, to communicate, to improvise, and to vitalize. It is trying to paint me into its camouflage. But I am not made of cubes; I am a beast, a vermin, a stray. There is wildness still coursing through my veins, synching me to life. It is from there that I hope to learn, not the digital beeping world.

I howl into attention so as not to drift into submission. In the way that a foul smell that is at first intolerable then fades. You get used to it and soon forget to hold your nose. Remind me tomorrow to be disgusted; I don't want to be subdued by creeping acceptance. Keep to the rigor of recognizing these digital domains as dungeons of the senses. Pinch me so pathways of perceiving do not drift into new normalcy—complacency. As creatures, we need analog time together; we need to do things that are not digitized; this cannot become a faded memory.

Day in and day out, I am cracking off the crystalized bits of culture that ossify around the edges of my perception. It is an endless job. Staying in line but not letting the line scribe me, staying in the grid but nurturing an untethered parallel self. Music helps. Dirt and forests help. Playing with dogs helps. Talking to people in other languages, making eye contact, dancing, making love, singing loudly in the car, reading poetry . . . these are the holes in the top of the jar.

It is not that the existing systems are corrupt. If they were, there would be human beings on the line that could be tempted by money, or fame, or sex, or aggrandizement of some sort. The humans on the line are not susceptible because they have been bound by policies. Every once in a while, it happens; we recognize a fellow creature. Sometimes, the entity in the chat box or the phone is named, but the name is false. The chatbot is nameless, grandchild-less, ancestor-less, senseless. The email addresses go to a vast void where "your call is important to us" is the spell cast to calm the indignation of being made into numerical paste.

I cannot find a way in. There is no way to sway loyalty. They are just doing their job. Nothing more, nothing less. It is not corruption but life-sucking, dehumanizing, fascistic, devitalizing horror. It is how things get done most efficiently for the company. It has been tested. Do not question; do not deviate: you cannot get the ticket dropped, the charge removed, the diagnosis reversed, or the policy changed. "Yes," they are trained to say, "I understand you are frustrated . . ." A company with no phone number is a terrible idea. A cashless society is a terrible idea. A society that has architectured a way to NOT receive complex information is in no way prepared to respond to the coming years of complexity in full bloom.

You see, this is not just a matter of temper tantrums and impatient self-entitlement. And while I feel a bit like Jack Nicholson raging like a troll in a fury, I know the adult thing to do here is to stay in the heat. Do not go quietly, wait patiently, or be polite; the adult thing to do now is to SCREAM. It is a matter of disrupting the algorithms that are wrapping tightly around us—like a digital boa constrictor binding

Combining

its prey. Nothing less than human interaction with life itself is at stake. Our breaths are becoming shallower; the constrictions are becoming comforting.

The relationship configured between citizenry and institutions forms the patterns of communication in a culture. The interactions within which life takes place, all day, every day, are within this meta set of narratives and normalcies. The seeds of system change lie here, in these seemingly mundane and banal communications with the larger structures of basic life. This form of intimate activism is released into the minutia of a moment's eye contact with another person who says, "STAY HUMAN WITH ME." It is a kind of demonstration that is not about taking to the streets but about responding through re-animating, lasting through the dull deprivation of departmental separations, and re-invigorating the hidden creature within.

To upheave the matrix is to do what it cannot do: meet each other face to face, person to person, breath to breath. Have long conversations about nothing important. Tell random stories, dance together, sit in silence, be there to witness birth and death. Analog contact is radical now; find it everywhere you can, from the inside, from the outside, from the sidelines. Rip the fabric of our notions of distanced departments of sociological control; this grid we are in is mind-fucking us into becoming zombies that have gone blank-eyed.

The ones who re-orient will likely be labeled pathological. The thresholds of normalcy do not include indulging the creature or coughing up hairballs. There has been a reversal; what is "normal" is not; what is "normal" is sick. How can it not be sick to disconnect from life? This is not how to live. Living requires humans talking to humans—finding ways to sort out problems with their combined creativity—learning together, off script, across multiple contexts. We need to connect if we are ever going to extoll the possibilities of interconnectedness; we are dependent upon multidimensional contact; we are, at core, analog. We sweat, we poop, we cry, we love.

Now is the time to veer toward a larger context, opening our perception into the transcontextual, where the idea of rain is felt as raindrops on your skin and the water rolling down your cheek—revealing relationships of another kind. We recognize the difference between the raindrops rolling down our cheeks and the tears of a friend who has lost someone. We can be moved to tears. The possibilities of interdependency rely on the skin-to-skin, the eye-to-eye; our fundamental dependencies rest and move in this contact; we are, at core, analog.

Need water in your home? Contact the department.
Need medical help? Contact the department.
Need tech support? Contact the department.
Need a break? There is a pill for that.

Wait your turn for the appropriate tentacle of the great gridded squid of Western society to grant you an audience. Or leave your number, and it will call you when it is ready to speak to you. "You will not lose your place in line if you choose this option." It is imperative that the urge to question this insanity not be quelled now. Re-membering ourselves in the ecosystems of life requires naming and debunking the notion that submission to the grid is in any way a good thing.

Have you noticed? The system is dismembering the membership of living organisms like me. The banks have sold the debt and robbed the world. The agriculture is poisoning the farmland and our children; the schools are preparing the next generation for participation in a world of fifty years ago; medical systems serve the pharmaceutical industry selling pills to treat the side effects of pills; law is not stopping ecological or human exploitation; politics is a farce; technology is in runaway warping every dinner table conversation and parent-child relationship through a tsunami of screens. It is broken. So, no, the justification for the grid is not holding. The system is not for our benefit; it is killing us and most of the biodiversity of the planet at the same time.

The beast must be let out of the cage. That last fragile thread of human wildness that remembers how to feel and express exasperation at the dull tones of impossibility with which the existing systems drained the life out of the world.

Who is best suited to begin this uncaging?
Not the people for whom the system is working.

Those who have been betrayed have met the cold demon face-on and lost. Those who have been given faulty court decisions and toxic medicine; those who are enslaved by their jobs working for less than the cost of living and did not have food for their children; those who have had their families broken by immigration officers and been caught in the nightmare of charges that fit the policy of the organization but do not fit life. Once the gloss of systemic trust is gone, you will be ready to re-find your roar.

Then, our relationships with the systems change. The curiosity, the courage, the care needed to activate awakens when there is no way to take the dehumanization anymore.

Nest in the arms of someone you love.
Listen to their small sounds with yours.
Try to find their scent along with your own;
heed the majesty in our remaining wildness.

COMBINING

I AM A CRAYON

I am a crayon clutched in hand ...
Smearing the colors of life
all over the table, the floor, the windows, the bed, the
sky, the stars, your toes,
My breath, our world ... and in the scribbles,
We are both set free —
From the clean lines of a metallic grid.
A thousand times again ...
We find ourselves to be so alive.
Nothing fits in the frame.

TIME IN WINTER IS UNDERGROUND

Sleeping in the radiance of the molten core.
Up top, the whispering days are frozen, crunchy, and foot-printed.

Making ancient mischief in the wilds of winter is a rebellion to the dulled heart. Here in the Scandinavian woods, where the old stories peek through the snow, we stomp and plow and scrape and tend to the tasks of the cold.

Still, winter is irresistible sugar fluff-tickles and icy surprising crisp.

A muffled quiet on the snowy forest floor, soft and soothing to slip a wandering thought into.

A hush in the crispness, lulling the trees to their root time.

Tiny minutes of golden light skip in and then hide again,
With a kissing promise that again we will dance with naked shoulders in warmth.

We dip back into the night by early afternoon.

For now, the candles, the logs, the blankets are all woolly reminders to remain in awe of the cold,

Old stories told ...
The season of filigree.

UNSILENT

Unspoken
Undefended
Unoffended
Underground
Untouchable
Unrecorded
Uncanned
Unseparated
Undressed
Unknown

Uniting.

MARROW

To maintain love, humor, and curiosity,
While simultaneously keeping grief in your fingertips,
Surrounded by inane exploitation,
To know the binds will have no resolve,
Requires salt.
It asks marrow of us.
Holding that damned disappointment,
And your warmth,
In the same hand.

Fig. 33. Vivien Leung. (2022). *Untitled*. [digital art].

Fig 34. Vivien Leung. (2023). *Bellflower*. [digital art].

New Blank Document

New Blank Document. The word processor application opens.

Is there space in the keyboard strokes for the pink of life?

Assembling letters into words into images that reach into feelings, inviting memories, things once known, regrets, and touching the purple bruises of this moment. Is there a chance now to say what we have known and previously tried not to voice?

People will demand clarity, accessibility, directness. They will tut-tut and brush off the brushstrokes that do not have the ring of sterility and know-how.

I apologize. I do not mean to be incomprehensible. I am not itching for cleverness or decorative prose. It is the bruising that is purple, not the words.

So much information is missing in the surgical extraction from context. It is inevitable that there will be consequences when that decontextualized information is the basis upon which decisions are made. There is no way to respond to the complexity around us without accessing our own complexity.

Thus, a variety of new textures of expression and comprehension are needed.

What you cannot hear now is the silence, then a flurry of drumming fingertips, erratic, stochastic. Time is peeking through thoughts in bursts of tapped-out words, in stewed mixtures of eras, some to come. Somewhere, there is a ghost of an old typewriter, a feather dipped in ink, stains on bark. Markings.

The weight in me says prose cannot hold the blood of this work. What to write? How to say what needs saying?

Perhaps the confusion frothing now in politics and culture is pushing the thresholds of recognition that more relational information is needed. Not more decontextualized information but more warm data. The warm data is what is between the stakeholders, between the organisms in an ecology, between the ideas, cultures, and languages.

Now, the fissures in understanding how to fit into a changing world have let in the smog of despair and a tsunami of pharmaceutical meds to soothe it. Perhaps that despair is proof enough to reveal the need to understand the missing vistas of interaction. Perhaps the innumerable efforts and projects that sought to do good fixing the world and failed … perhaps those are evidence enough that the way of comprehending society, economics, nature, each other, sex, food, education … was not enough to actually meet the mess where it is forged in ideas of control.

I learned recently of a project by SEED to "End poverty one person at a time"—and thought, oh my … why did they ever think that poverty could be anything less than communal, cross-cultural, international, transcontextual? Millions went into that project, and it was a disaster, of course.

None of what we are faced with can be tackled in isolation. Working with parts and wholes is not so helpful in this case. Zooming in and zooming out is insufficient; we have to do both at once. I think of the systemic combining as a kind of broth instead of linked pieces. The difference is akin to the contrast between that which is interconnected and that which is interdependent. Interconnected things can be taken apart and fiddled with, fixed, and replaced. You cannot take the salt out of the broth or the grapes out of the wine.

The connective tissue repair, the mending of rips in perception—this kind of thing is not accessible in direct language.

Warm data requires more of you. More than your job, your expertise, your title, more than you know you have to give.

You won't find it in the graphs,
Or the stats, Or the lists of stakeholders.

The gaps between the subjects, objects, verbs, and grammar of tight capsules are a vast lost space where the most important sensible assessments can drain out.

Have you ever filled a jar with marbles? It was not full. The space between them could still hold what seemed like a full jar of water. So many isolated institutions, people, ideas, societies—and adjusting them does not seem to fill the jar.

It turns out that systems change is more profound than tweaking a few institutional protocols. It is more profound than branding green goods. It turns out that systems change is in the way my metabolism looks forward to coffee.

If that is the project, then that is where the work is … but it must be done carefully, very, very carefully. In that tender liminal realm, much resides—the ability to love, the capacity to desire another relationship with each other and the world, a relationship that is vital, not exploitative. Status is re-sculpted there.

Combining

Let's go there—where everything depends on everything in infinite responsiveness.

What is the point in averting my eyes from this very obvious realness? Is it inconvenient? Yes, but only in so much as it will rub against the expectations of planners and funders in ways that have no existing protocol. Right now, the broth is no one's responsibility. It is not the educators, or the politicians, or the doctors, or the lawyers, or the scientists, or the business people, or economists, or the artists (maybe the artists hold a little bit of responsibility)—and it is suicidal for everyone to keep doing the jobs they have now. To stay alive in a changing world is to create another way of life.

Do you remember stick bugs? They look so much like twigs they are hard to find in a tree. Ask yourself, how did that information of how to look like a tree get inside the insect? Information about tree color, texture, proportions of twigs to little branchings—somehow became the information that forms these creatures. They are shaped by their surrounding organisms. What if you were a stick bug in a tree or a snow fox in a tundra? You are an extension of the contexts of your life, and they are an extension of you. You do not end with your skin or your tax ID number. You are generations, communities, you are ecologies. Who am I as a stick bug if there is no tree? Who am I as a snow fox if there is no snowy landscape? Wondering why systems change is so tricky? Wondering why there has been so little shift after decades of discourse on how humanity needs to respect the environment, limit growth, stop exploitation?

Living differently is no joke. It is not a refurbishing or a greener renovation of my current patterns and habits.

Into the landscape where none of the maps apply—the deep down. Where your health is the health of the next generation, the health of the community, the health of the biosphere . . . it is not a doctor's office visit, a diet, or an exercise regime. Your health is a measurement not of your vitals but of your ability to perceive and give vitality to the overlapping living processes around you, beyond you, within you.

The transcontextual work is there, waiting, aching for the tourniquet of our separated perceptions to be untied so the blood can flow. The grants and funders have yet to hear the frequency. The politicians are quick to play the chords of emotion but do not know what they are tampering with.

However, it will not be in the direct language of strategy and authorized leadership. It is in the intense generosity of contact and overlap. It is in the wash of emotion and irrational mythological hangovers. The sticky plasmas of life will make the interdependency of the work ahead less messy.

The assembly of words, ideas, notions, and voices combine to form and inform. But they are just words, and some will lead us to lostness.

Save file. Replace file. Rename file. Find.

YES

In the flush of your warmth – yes.
In the breath between your laughs – yes.
In the unscripted and unfound – yes.
In the shattered and betrayed – yes.
In the minute particulars – yes.

COMBINING

DIVIDED WE FALL TOGETHER

In a thousand colors and textures -
an arc reaches
into the notion that there is someone
who will help you,
and you will help someone,
in times of confusion.

When the emergence
has lifted the veil
and the betrayal is revealed.
I need you,
and you need me.

Un-clocking from accelerated anxiety,
from the loneliness of independence
into interdependence.
On a bending beam
over the sharps of
cracking confusion

There is a path to you.

Come here
where I can hold you
as you hold me
just in time.

FOR YOU

To catch a smirk in the turmoil.
Because if you don't eat paradoxes at
breakfast,
What do you eat?
And if you walk on only two feet,
Your balance is lonely.
Irritation is the gut saying ...
Forgodsakes! This is about life!
While the storm rages above and below –
Without and within –
Let it blow and flood.
Attend the tiny.
And the small curls of breeze
Are the next folds
Of the ecology
Of you.

HOW DO YOU PACK?

I do not long for how things were before.

Nor do I long for some fabricated future utopia.

I do not long for things or money, or status.

I do not long for more time or even for fixing the broken corners of my heart.

I feel the up-ending of so much that has needed to be upended for so long ... but I do not long for the chaos.

It comes like a fever to a sick body.

I do not know what a healthy system would be like, I do not know how to long for it.

How do you pack when you have no idea where you are going?

SACRED COMMUNICATION

Some things that seem like they should be said cannot be said. Not because there is a law or a rule but because the ecology of the situation will not accommodate particular forms of communication. People cannot speak up sometimes because it is impossible for them to speak up. That is not because they are inadequate; it is because the context has closed in. The complicit agreements about what's possible to say are sometimes necessary and sometimes deadly.

There are some circumstances where I would let someone know they had spinach in their teeth. Other circumstances where that communication is not going to happen. On a more serious note, this limiting of communication is built into most relationships that have abuse in them. The silencing is not voiced as a rule; it is implied and upheld in other ways. So, there is never a transcript to turn to. They never said it. They did not have to. No one spoke up, they could not do so.

How do these limits shift?

They do not do so on demand. Something loosens. Something strengthens. Usually, some other context provides new information or sense of being that allows for something new to happen.

The transcript has so little of the information on it. The recording too.

If the world is made of relationshipping, then communicating is the sacred sauce of vitality.

Cheating in communication, to win, to take, to humiliate, to gotcha, to gaslight, to decontextualize, to trick or manipulate ... these things might be the most sinister de-vitalizing practices.

An Ecology of Assholes

The world is a beautiful place, full of souls who want only to be loved. Humanity has achieved wondrous feats of elegance, humor, grace, and poetic creativity... but there is also the asshole factor. By asshole, I mean jerk—I don't mean murderer. Certainly, murderers are assholes, but not all assholes are murderers. With all the name-calling and finger-pointing right now, why not take a minute to apply the axioms of systems thinking and ecological patterning to something closer to home than saving the world? We all know a few assholes, and those who can admit it might confess to having even occasionally joined in on the assholery. That much is a given.

Commonly and paradoxically, assholes are thought of as individuals. This is a mistake that should be untangled. For that reason, this piece is a short exploration into the way in which assholes coexist within frames larger than their own sphincters, even though they may not realize it. Perhaps, this is a moment to look at an ecology of assholes. It is all very well to say we are all interconnected, but what about the implication of being interconnected to all the assholes? And what does that say about the non-assholes? Are we all in the oneness? Oh, no. An "ecology" has a couple of important characteristics that we would do well to keep in mind for our analysis. First, as Merriam-Webster says, ecologies are found in:

"The totality or pattern of relations between organisms and their environment. (n.d.).

Is the asshole really an isolated island of their own jackassish-ness? Or is their relational interaction taking place within the larger context of communication that the asshole is responding to? Is the asshole-ness within them? Or is it in their relationship with either you or the world?

People prone to humiliating others, lying, or displaying the arrogance that comes when they believe their own life to be more valuable than others cause those around them harm, leading to a loss of faith in humanity. They cannot be trusted; they judge others; they are blamers; they hold themselves to be "right" and clever, while others are stupid; they boast about the way they've shamed someone else or made them suffer—they go on and on about how others are jerks.

Ummm, wait a minute... am I an asshole for making this list? For judging, blaming, etc.? Self-reflection is not generally a quality of an asshole. (Or maybe it takes one to know one?)

In other words, when you ask, "Who do those assholes think they are?" You might confirm your suspicions by listing their characteristics: personality, politics, profession, family, nationality, and so on. You can then point to them, personally and individually, and say with some evidence in hand, "An asshole is an asshole." In that sense, the asshole is indeed an individual with a cluster of circumstances resulting in their becoming a douchebag.

Part one of the paradox is that, in this way, the assessment is correct. The way in which that person experiences the world is uniquely their own. Their family, their culture, their job, their friends, their acquaintances, their sleep habits, their microbiome, and so on—holistically speaking, all of those things come together to form the filters through which that person experiences the world are uniquely theirs and no one will ever be the same as they are. No one will experience the color blue as they do or see the same meaning in a poem. (Maybe assholes don't read a lot of poetry... or maybe they do.) They are their own lens, and no one else has the same one.

But how did they get that way? Part two of the paradox asks if any aspect of them is not influenced by their family, culture, language, food, etc. Is there a definable part of them that is outside of the great interconnectedness? In this sense, they are a combining of all that they embody. What learning took place in their world that contributed to their assholing? Is it really a choice to be an asshole?

I am not suggesting solving this paradox. Living within the interconnectedness of assholes is not something we can opt out of. To be an asshole is both a choice and it is not. Even as non-assholes (or so we might hope), we are all caught in a web of deplorables, and in that sense, we are part of the systemic ass-hatting of our world. The next ecological characteristic is interdependency. Ecologies are relational and interdependent contexts. There is nothing outside of the processes that are continually forming and informing the ecology. Assholes are not stand-alone entities.

So, maybe the nice people are really the assholes because they go around pointing out assholes to make themselves look good? Ever wonder? Is every asshole so wrong when asking, "How is this person trying to screw me over?" Perhaps we all even need a little asshole-ness to keep from being pushovers?

Or is the asshole identifiable as the one who is constantly pulling things out of context and dissing them? It is the ultimate violence to take one tiny piece of information out of a more extensive set of conditions and circumstances, decontextualize it—cast it as the TRUTH, and then disavow all other contextual input as "beside the point?" That is certainly what assholes do. And they do it to people as a form of disrespect—but also to other living systems, art, ideas, other people's projects, and so on.

Assholes don't get interdependency. They don't get that they are in interdependency.

But, then I find that I don't get how it is that they don't get the interdependent consequences... and in that swift move, I become the asshole. Within this dreaded reflection, I see it is me who does not ask about the ecology we share. It is me who cut the picture and cleaved the context.

The real problem with assholes is that humiliation, disrespect, and decontextualized, judgmental arrogance contaminate the ecology of our communities. The overtones of life, in general, can go sour when vile, exploitative attitudes abound. Assholes underestimate the profound awe of each remarkable living being. In doing so, they escalate trouble untold. Fair enough, you may say, life is a bitch... but just keep in mind that it takes a great deal of collective tenderness to heal ecologies, asshole.

THE CRINGE

When I look back at times in my life when I really screwed up, it is the deep cringe, the gut level embarrassment and the contrast to how I have shifted thinking/behaving since then that are indicators of having really LEARNED.

Doubling down on how I did the best I could at the time is rather mealy and insufficient. No matter how true.

The contexts at the time were in play no doubt, but then ... there was learning. No point in trying to defend the idiocy of the past. Humor is a better healer than blame or denial.

We are in a time of transformational cultural learning. So, there are many, many, many things to look back on and cringe to think of now. Let the cringing show the learning.

Learning together.

A thing I can do:

Risk ruining the dinner party.
Risk being a difficult friend.
Risk breaking the silence.
Risk learning about the cringe things I thought were okay to say but weren't.
These are tiny things ...
They are the least I can do.
There is much more too.

REJECTION

Hello.
Today there is a chance.
Actually, many chances - to notice the invisible pathways of reductionism that continue to generate rejection.
And what does rejection generate? More hold back, more rejection, reductionism, despair, & unseen-ness.

So ... Hello.
To the possible giving of mutual dignity.

How was it to be the one rejected? Remember?
How did it feel?
Socially, professionally, intimately.
Did you feel like lashing out? Shutting out? Did it feel unfair?

And the grounds on which the rejection was justified ... How did those grounds feel under your feet?
The experience of being underestimated, reduced, & decontextualized -
These are conditions for the most dangerous weapon. Exclusion.
Rejection has long & deep consequences.
It is a poison that generates more poison.
It jumps into other contexts.
A lot of attention is needed to tend the way rejection is dished out.

Rejection comes at a high cost, sometimes lasting generations.
There is no easy method to un-reject. The opposite of rejection is not acceptance but dignity. The rejector forgets their own experience of being rejected.
And it must be, at times, that a rejection is necessary.
The who & how of rejecting others will divide us all.
Beware of the metrics of rejection.
I greet you.
Hello.

TWO BAD QUESTIONS

Here are two bad questions.

Is the future one of authoritarian control of our lives to protect us and the environment from humanity?

Or is the future one of unmitigated greed and corporate exploitation and destruction?

Neither of the above questions is asked from within an ecological perception. Both are infected with the same error of industrial control.

SOMETHING NEW

An arrangement of notes and soundscapes in an unnamed
music –
I cannot quite find where it meets me.
A combining of spices I do not recognize stews into an
unknown taste.
I am on a road, but I do not know the landscape,
the turns,
the hills.
I do not remember who once lived in that house over
there.
I am not of this.
And it is necessary to pay attention.
To home in,
On a weird combination of remorse and curiosity –
A chemical unbinder of predictable nexts –
An emulsifier of new nuance.
That allows me to be honest about the nasty history,
Honest and still lost.
In the unfamiliar –
It is finding that door to a world that does not track
on old rails,
Opens only with pause.
Slams with eager practicality.
In the future I will be different
And the me of now will be a foreigner –
An awkward guest in my own history.
Change is when I look back and see I did not see.
And know that I still don't know.

Swerving

The work of the coming decades is not the work of manufacturing, of software development, or of retail seduction; it is the work of caring—for each other and the biosphere. In that care, there is the hope of finding new ways of making sense of our own vitality. The "my" in my health is not mine; rather, it is a consequence of my microbiome, my family, my community, and the biosphere being cared for. The work ahead is not clear or clean. It requires intense integrity, patience in ambiguity, fierce dedication, raw vulnerability, and bleeding humility.

There are sufferings now. Non-feeling is a non-option.

But, it is different than before. Fissures are forming in the idea of fitting in.

One suffering is the suffering of being incompatible with the going game, where the grooves and pathways of life others presume normal are impossible, uncrackable, and untenable. The isolation of knowing there is a frequency that others can hear and not finding it. It is like the instructions are in another language, written in invisible ink. The realm that others occupy—comparing successes, jobs, credibility, and status—is mostly abstract, out of reach, and to fit into it requires extreme inner acrobatics.

Another suffering is that of fitting in. This is the suffering of finding oneself so synchronized into the contextual pattern structures that it is impossible to perceive, shift, or change the habituated systems of the day-to-day. This is the suffering of successful compatibility in the dominant socio-cultural assumptions, such that every move seems to feed the monster of current institutional confusion. This stuck-ness is waterproof to knowing the deadly consequences of not changing—both to humanity and countless other organisms. Like a Kafka story, the visceral experience is of recognizing that the logic of the surrounding systems has consumed play and rebellion. You will lose your job, your status, and your credibility if you swerve.

One suffering is the suffering of loving someone who is suffering.

This is a terrible time to aspire to obsolete normalcy—whatever that is—when the mandate for ecological survival is contingent on breaking from the sense-making that is entrained. But, to orbit outside the flows of the current systems of modern life is to be excruciatingly isolated. It does not serve to facilitate numbness to this pain. It hurts because it hurts. The tears are cultural, conceptual, and ecological.

The double bind of this era is that the continuance of our species requires discontinuance of current means of survival. Business-as-usual is a swift endgame. Yet the rent must be paid, and breakfast must be possible. To live through next week is to take part in systems that are destructive to the future.

To get unbroken, a breakthrough is needed.

There is need.

Those who are in step and rhythm with the way things are, despite future incompatibility with life,
Those who read the signals and follow the signposts in the map of today,
Those who have fitted infinite capacity for contextual response into one frame of limited contexts . . .
Need.
Those who are hearing another tonality,
Those who make sense in another way,
Those who do not fit . . .
What does it even mean to fit into a rapidly changing world?

Any small window of another sensory experience is more precious than gold now. It is time to listen carefully. It is time to pay attention in wide ways. Let logic that excludes variables unravel into warm complexity. The fodder for this mutual learning is connections connecting in unexpected ways, discovering flavors of thought—mapping textures of knowing.

Un-labeling each other is the greatest rigor and the greatest gift.
Allowing multitudes of selves to mingle and form new ecologies of communication.
Abandoning the flatness of analysis that prides itself on non-emotional rationality. If the interaction is not funny, angry, curious, confused, indignant, and at least a little bit destructive . . . it is not worth ten minutes now.

What is healthy or not healthy in a changing world? What is it to know yourself in a culture that is unweaving itself? Who are we now? I am resisting the antiseptic distance and diving into relationships of mutual learning. Relationships in which there is an acknowledgment that it is a violence to reduce ourselves and each other to definitions, titles, and labels. I am not gone or fragmented; I am real, and confused, and unscripted. As such, I am no source of tricks or easy methodologies. I am not interested in technique; it obscures the unsearched for complexity. Rather, I am a sea anemone, all tentacles sensing into our combined vitalities and learnings, exploring our mutual dignity, noticing paradoxes and contradictions, tones and strangenesses. There—in the warm data of our interactions—is where entirely unanticipated possibilities are to be found.

Combining

A LETTER TO MY IMAGINATION

Dear Everything,

I am writing to ask a favor.

The weave and weft of this imagined world has grown into a tent with no door. All around me the hearts of others are longing for another imagined world, summoning, without realizing it, the imagination that is so actively illustrating the current version, to redraw, re-sense, re-vision.

The vision is an imagination that is convincing itself that it can change itself. The patterns are repeating in the name of change. The favor I am asking is rhythmic.

Can you send a pulse? A song perhaps to weave into anew? New gaps to make connections between?

Or would you like to visit my hands, these same hands that 15000 years ago spent their days in another imagined world. Back then my ancestors' hands held rocks, held wood, held dirt, and made them into cosmologies of story. My hands have known many imagined worlds where my memory imagines that this current installation is the IS. Always through imagined notions have we known the world we share, not how it is. My hands remember many stories of how life is. None of them were real, but then that isn't necessary.

Can you tease my hands into the shape of another remembered world, one that reaches before the time of houses with right angles, one into which there was another version of time? Can you rip a tiny snag in the fabric of knowing to remind me that it is only and always imagined?

It is not that I am nostalgic, no. It is that the air in this imagined world smells like numbness, and certainty. The tightness of the stitchery has made it so difficult to place life that makes life before life that makes the current collections of illusions continue. It is time to be roused from this dream into another.

Living is imagining life. The question I am carrying is, "Can I keep in mind that it is imagined?" Can I perceive that my perception is only that, a wonderful rich set of imaginings - from language to government, from culture to history, from the notion of the future to what happens after I am gone? I am only imagining.

There is much imagining of imagining of the future going on. As though the imagination of the future would not be presoaked in the past, a linear casting of vision - seasoned in now ... with such precision that it calls itself change. A great prank has been played. I promise never to re-imagine the future. I leave that to the underground, unseen, un-named realms. Better not to narrow the path. Do you see how convincing this version has become?

I am asking for a favor. Please poke your nose through the threads, rip the seams, dance on the knots. I will know it is you by your shaky breath of incoherence. The way you sneak in through the mole holes in my garden of ideas. The way a fish grows a new fin, the way a plant responds to a new climate, the way a baby meets its world by wrapping small fat fingers around the first thing possible and tastes it, making a memory farm.

I am calling you from the me that is beyond me, the "my" imagination that includes the far reaches of self to include life. Please don't send the imagination ambassador from the department of human assumptions, send the other one who lives in the creative belly of evolution.

If at first I do not welcome you upon your arrival, I beg for your patience. The dream seems so real, it may fool me. Hold my hand - the same hand that planted seeds and built fires, that made food and music, that held cheeks the of children, the hearts of the wounded, and fabricated a clicking, beeping realm of machines. Hold my hand, perhaps my hands will remember you.

Love,
Nora

Fig. 35. Rachel Hentsch. (2023). *Combining Elements*. [photocollage & digital art].

Fig. 36. Rachel Hentsch. (2023). *Water Squares.* [photocollage & digital art]

COMBINING

What does it mean to take a stand in an interdependent world?

Liminal Leadership

> "The past is our definition. We may strive with good reason to escape it, or to escape what is bad in it. But we will escape it only by adding something better to it."
> ~ Wendell Berry

Reason has become unreasonable.
Things are changing,
And the reasonable path of maintaining life as it is today,
May well be one of short-sightedness and destruction.
A new frame of reason is needed.

This is an unreasonable chapter. I am becoming less fixed on presenting a rational strategy, despite accusations of idealism.

The future is coming too fast and shimmers with patterns that are, for now, still invisible. Going forward, who knows what will happen. People are nervous. Consequently, leadership as a topic is popping up everywhere. I am not confident "leading" is the best metaphor for what is needed now. But for lack of another symbol, let's continue with it.

Inter-systemic Change

Whatever leadership used to be—it used to be. Now, it has to be something different. Now, we all have to be more than we were. Leadership models come in many flavors: strategic leadership, leadership from behind, organizational, innovative, creative leadership, collective leadership, transformational leadership, cross-cultural leadership, team leadership—the list goes on. However, the kind of leadership that I want to explore may not be identifiable as leadership at all. I am interested in a kind of mutually alert care and attention to the well-being of all people and ecological systems. This kind of leadership cannot be found in individuals but rather between them. It cannot be found in organizations, nations, religions, or institutions but rather between them. I have called it "Liminal Leadership" to highlight the relational characteristics.

Inter-systemic change is at hand. More than change, and more than system change, the interdependency between systems of economy, health, politics, ecology, and communication is where the change lies.

Combining

This is a murky territory of alive in-betweenness. The interdependency we are discussing should not be thought of as a part that can be replaced in an engine. It is elusively not in the economy or the education system, it is not in politics or the health system, it is not in the media, or even the culture—it is in the way in which these aspects of our world are steeped together in a slow-cooked stew. The ingredients of the socio-economic stew cannot now be pulled out, but the chemistry can be tended.

As citizens and human beings, we cannot point to these institutions as "them"—there is no them. All of these contexts of society (and more) are in a kind of ecology of interdependency, pattern, and relationship. You and I are simultaneously in the systems and occupying the position of observer or change-maker. We cannot get out.

But, we are also within another ecology: the ecology of the biosphere. The difficulty is that the ecology of our institutions does not support the larger ecology of the Earth's systems, nor does it support the patterns of natural processes in complex living organisms. With the exception of a few remaining peoples who live deep in the natural wilderness, most of humanity lives in the middle, in the liminal space, needing both ecologies to survive. Needs, like breathing, eating, loving, and making community, are arguably impossible to change, whereas rethinking the structure of society is merely extremely difficult. Keep in mind the "reality" of these socio-economic systems is a human construct; the deer and the sea algae do not buy food.

As our systems begin to fray in this unraveling time, reorganization is necessary. Who will lead the way? Who are the experts at being in the liminal space? Who are the professionals who know this territory in which each day is touched with health, economy, media, politics, education, and the earth? … Who? Of course, the answer is all of us.

Vignette One: *When I moved to Stockholm, Sweden, I spent my days on the arm of my best friend. He took me everywhere. This city was his city; his body was extended into it. So, I never paid any attention to where we were going. I leaned into his knowing and merrily went wherever we went. Alone, I fumbled with maps and place names I could not say. The effort was not merry, and the results were patchy. I travel constantly, and I don't have this issue in other cities. I deferred my sense-making of the city for plenty of good reasons, but in doing so, I habituated a lack of attention to the landscape. Likewise, leadership has been located inside "things," organizations, nations, departments, ministries, and authorities—and people have leaned into that leadership. Leaning into the liminal is unfamiliar.*

Noticing Holdover Ideas of Collaboration and Competition

I recognize that it is believed that great leaders have brought us to this point in history. I am willing to acknowledge that perhaps during the previous phases of human and societal evolution, the notion of "leader" was necessary. But now, to meet the complexity ahead and make the evolutionary jump to co-existence, I think we need something different. Leadership has an ugly side. The image contains hubris and the bloodshed of the colonial conquerors. It is laced with competition, ambition, dominance, and arrogance.

The notion of leadership pulls the focus to individuals and away from the contextual conditions that made them. Is the tree tall because it grew more cleverly than the other trees? Or is it because the soil, light, water, and biodiversity of that particular acorn were nourished to provide the conditions for thriving trees? It is precisely this contextual, relational process that the future depends upon.

I was sent an entry for a contest to win a prize of several million dollars awarded to the person that provided the winning project for developing social collaboration. The irony was too much. Competing for ideas on collaboration is a perfect illustration of why so much change-making today is riddled with toxic fumes of the last century's hero envy. The rush-for-the-gold and step-over-the-next-guy approach is the thinking that got us into this mess. Surely, it won't get us out. This sort of contest will divide people who could be working together; they will need to keep secrets about their work lest their ideas be stolen; they will claim credit instead of giving it. The hunger for the keynote gig, the best-seller, and the viral meme about how to save the world is better suited for Wall Street than for change-makers. Watch out. The future lies in the capacity to understand and respond to interdependency. When you see collaboration with a promise of fame or awards, you are seeing lingering ideologies of capitalism.

The change-making ideas that hold onto the realism of last century bring the baggage of mechanistic thinking, capitalistic exploitation, and dog-eat-dog aspirations with them. At first, it is hard to see how these holdover poisons sneak in, but they do. In the form of a boxful of visions for new economies, there is still the issue of how to produce wealth without exploitation. In the greening of consumer goods, there is still the need to fuel the furnace of desire to buy more stuff. Integrity is being reduced to a market value, and making sense and finding meaning will soon be for sale.

Collaboration is an idea unconsciously attached to the mechanistic world in which many parts are assembled to create function. But in living systems, collaboration is much more than each doing their part. Collaboration is the readiness to show up and do what needs to be done in improvisation and mutual learning. In the liminal realm, our conditions are co-generated. This is an important qualitative shift of interaction that translates roughly into: idealists are more needed than assholes.

GIVE CREDIT, WORK TOGETHER, LEARN TOGETHER ACROSS FIELDS, ACROSS GENERATIONS, ACROSS CULTURES, ACROSS SPECIES

Complexity and interdependency are messy. The relational processes at work are in motion, always calibrating, changing, and compensating. In between the hours, in between the phases of evolution, in between being professionals and parents and lovers, and friends and patients and citizens, and activists and athletes.... In the liminal land of being alive together in this incoherent moment, there is mutual learning. Between us is the genesis of the ability to perceive and respond to the complexity of this time.

I will meet you there—in the liminal plaza of our shared future. I know how misty that may sound.

The tone of skepticism that sterilizes the complexity of the way things ARE is toxic to the vulnerable visionary seedlings. When new ideas appear, their inventors often get their heads chopped off, even though leadership is supposed to be about discovery. It is about doing new things in new ways. If (and it is a big "if") humanity learns to live in a new way, I believe we will do so by learning together. This will not be because a hot-shot author has a new best-selling book on change-making, a viral meme, or a super TED talk.

Our liminal leadership will be as people together in a struggling biosphere—just you and me and the other seven billion mothers and daughters and fathers and sons. We will not lead on behalf of a company or a nation, not on behalf of a religion or a belief system. We will hold each other through the storms of economic volatility, ecological turmoil, and political insanity. There will be trauma, pain, and loss through which our solace during this transformation will be nothing less than the creative expression of tenderness. Healing together is learning together is leading together. Together, it includes the human and non-human world.

Whiplash

So many lines have been drawn between clusters of people now that it is difficult to keep track. They are overlapping in a chaotic mess of loyalties; the confusion is profound. Witnessing this moment is gruesomely fascinating. Separating one group from another is a bad habit that, once started, can go on forever, with generations of damage in its wake.

People can always point their fingers at another race, another generation, another class, another level of education, another religion, another profession, another gender, another health condition, another nationality, another political party, another lifestyle, another philosophy, another body type, another culture, another language, another blood type.... And the list can go on forever.

It seems to me that as the illusions of our system crumble, each grouping of ideologies ossifies in its own particular frequency and becomes less able to hear the others. The sense-making apparatus of our culture is losing its grip. The lines are moving. It is as though the gladiators of opposing belief systems are twisting their necks. Disoriented, these warriors are not sure anymore which polarity is which and where the binaries are. The fabric of left and right political parties is wearing thin, and economic inequality continues to rise. While it is terrifying to see the humiliation between groups, the fact of so much rapid change in the system is perhaps a sign that movement has begun. But in which direction?

History tells us what humanity is capable of, but only in part. There is more to find out about who the two-leggeds are. We have not come to the end of the story. (Or have we?) Are human beings naturally aggressive and greedy? Are you willing to naturalize those behaviors and cast them in the concrete of our future? I am not.

I don't know what humanity has hidden in our inner wilderness. I am willing to hold the door open since the alternative is just too dark. But short of a fundamental reorganizing of embedded assumptions of life and being alive, humanity may not make it. So, are we ready?

THE FUTURE CAN SEE YOU NAKED

Most of what matters now won't matter later. Coming generations will shake their heads at the sacrifices their ancestors made for material wealth. They will not care how much prestige you gathered, how many bitcoins you bought, or who considered you famous, or even what widget or vaccination you invented. If humanity makes it to the next level in the evolutionary game, it will be through recognition of our interdependency with each other and the organisms of our biosphere.

If you consider this moment from the vantage point of thirty years from now, there is nothing to hide behind. The excuses of dutifully perpetuating this destruction from one day to the next won't hold water (literally). Remember the Nazis at Nuremberg who said, "I was just doing my job."

Advocating for the delicate ecologies of life and humanity is both an active and a contemplative practice. Protection goes meta. Protection of me becomes the protection of you, and protection of us includes protection of the ecology in which we both breathe. This interdependency is what gives every living system its vitality. It is the most ubiquitous experience of every living organism. And yet, it is not mentioned in the United Nations or the Constitution of the US. It does not have a Wikipedia page. It remains, it obtains, and it continues regardless of not being recognized. The interdepending keeps interdepending just out of our hearing range, just out of our color spectra, just beyond the horizon of our logic.

THE SENSORIAL POTENTIAL OF HUMANITY IS LARGELY UNFOUND

Is it possible to see what we have not seen before? To say what we have not said before? To love those we have not loved before? To think what we have not thought before? To live in ways we have not lived before?

Look back into history, and you will see what people thought they saw and how they thought about what they saw. But, looking forward into the coming decades, perception and sense-making are wide open—unless they are already slammed shut. Numbness is dangerous now.

Authorization will not be granted to enter new realms of sensorial experience and description. No one will validate, no one can substantiate, or accredit that which has yet to be perceived. To show up now is to show up with one's whole self, body, intellect, emotion, finance, career, and family—and to show up ready to learn. The issues in this historic time are complex, and it takes complexity to perceive complexity.

Fresh skin
feels different sensations
With incoherent language, new sentences arrive
on unfound tenderness
breaking the membrane of numbness.

Three Things to Remember About Responding to Interdependency

1. Wicked problems require inter-systemic change—not siloed solutions.
2. Taking action before perception change produces repeated errors and short circuits the necessary complexity. Ditch linear strategy.
3. Perception is intellectual, emotional, physical, cultural, and relational. Making sense is sensorial. Increasing sensitivity is necessary to find new ways through old patterns.

Are we ready? We better be—because increasing sensitivity is an opening to also feeling the pain of so much exploitation... That pain asks a question—can I bear the tenderness that real systems change requires? Years, decades, and more than a century have passed in which brilliant minds with breaking hearts have tried to create change in the institutions that frame our lives. They tried incrementally changing the system from within. They tried using the legal system to change the laws. They tried becoming politicians, teachers, doctors... but the institutions did not budge. The multifaceted crises the world faces today are proof enough that the establishment is not built to question itself. The pillars of civilization are pinned under the stone slab of the last several centuries of assumptions. Pillars of politics and money, education and medicine, psychology and religion—structure is hard and hard to change. The institutions have no water in their edges, no improvisation in their memory.

Vignette Two: My grandfather, William Bateson, gave up on what he referred to as "the establishment" in about 1908. He coined the term "genetics" and watched in horror as his work was taken by the science community to prove eugenics, and then as the scientific work was taken by the politicians toward creating platforms of exclusion based on race, and then by the journalists who used the politics and the science as propaganda, and by the academy to facilitate the ambition of eager department heads. He would have none of any of it. I am still catching up. I have been quicker than most and slower than I should be to really, fully, completely admit that "There is no 'there' there." I wanted to think that there was a possibility of incremental change, even a little. And that democracy, even though it is not perfect, would be able to improve with the hard work and sparkling minds that have walked through the past decades. But, increasingly, I am seeing that a refocus is needed. The institutions are bound to their continuance, bound to each other, and bound to crumble.

There Is No Glory or Profit Whatsoever in Liminal Leadership

Vignette Three: Being a parent is sometimes dangerously close to playing God with someone else's life. I was afraid to send my son to a professional acting school when he was fifteen. He was a good student on a path to a good university. I asked his acting teacher if he thought it would curse my son's life to send him to study a skill that would likely land him a lifetime of waiting tables, and his reply was this: "If your son wants to go to acting school, don't send him. If he will die if he doesn't act, send him."

Anyone who wants to help usher in a new way of living that honors the well-being of all people and other organisms had better be willing to risk everything to get there. It will take nothing less. There is nowhere to hide.

Embracing complexity requires the integrity of having gone through the dark night and knowing that while you may not have a plan to face the confusion, you will show up completely. I don't know about us.

Are we ready to risk everything? What would you do for someone you did not know, for a forest you've never entered, for a future you won't be here for? Would you re-evaluate the ownership of your stuff? Would you let go of your position? Would you show up nameless, penniless, and invisible to the collective and offer your bare hands? Would you be willing to find dignity and joy in the care of others who may not at first be kind to you? Will you prepare for the crises ahead by building a bunker full of everything you might need, or will you prepare by readying yourself to help others in need?

Do you need to wait until there is an emergency to be activated? Or is now okay?

Can you walk in your integrity? And can you help pull the broken glass from the souls of the idealists who kept the door open?

---

* Previously published in *Kosmos Journal*.

While i am system i am also more.
The system isn't going to change the system.
(and while i am the system i am also more systems)

Words to be Careful With

There are not enough words to express the liminal contextual spaces in which the crises that must be addressed are combining. While it is natural to use familiar words, I am increasingly hoping that others know that I don't mean them in the way they once were meant. But, that is a slippery slope. Caution is helpful.

Language is important to the processes of perception. Through language, communication and communal sense-making are sewn. The jargon and go-to concepts of a community's thinking are formed and informed by the words used. For this reason, I am compiling the beginnings of a list of words that I have begun to use with increasing care. For many years, I have been interested in language that is used around change. I have become increasingly sensitive to language that tags into old thinking patterns, the possibility of perceiving in new ways can be stunted.

Sometimes, it is necessary to make up new words. I have done so with "Aphanipoiesis," "Symmathesy," and "Warm Data." These terms are my attempts to increase our sensory perception and capacity to express vital relational processes. I have also brought the term "transcontextual" out of hiding and used it to replace the word transdisciplinary when I am not speaking directly of academic structures. But if the language used to describe the interdependency of our work is too foreign, it will wash over the surface of other people's understanding. It won't stick. And at the same time, if the language is too familiar, it will pull old mindsets into the new work and trick us into a false sense of change. There is also the distinct possibility that new words will get metabolized by old conceptual habits so that the itch to innovate is scratched while the inertia of the status quo prevails. That won't do.

In this chapter, I have listed some words that are familiar to my colleagues who are dedicating their lives to the changes in current systems. My point in calling attention to this vocabulary is to insert a pause of consideration before use. I am not averse to the use of these words, but I am convinced that a more attentive use of them will better serve our communication needs. I hope this list is of some help, and I invite you to add to the list. As we move into the coming decades, I have no doubt that there will be changes in the language used to communicate the changes, and the changing changes . . . (what is change anyway?) For me, it is okay to retire some words that were once useful but now carry too much baggage. Ultimately, meaning is made between us and is largely non-verbal anyway. The words used are only important in so much as we keep an eye on the slippery way they can drag us back.

Nora's List of Words to be Careful With:

Agency

How can we be careful this word does not denote an objectified, instrumentalized notion of action? Why does it matter? Because it blocks the perception of contextual interdependency. How can any organism do anything not informed by and existing within the communication and relational process of its context? The possibilities for perceiving any action are within the contextual limits of the organism.

> The concept of "agency"—the idea that "persons" are able to "initiate their own movements"—and that such "capacities" originate from "within the person" are not processes that are observable. Instead, these ideas are part of the preunderstandings that we bring to any "observation" that we make of a psychological act. They are part of the intersubjective conceptual backdrop that we develop as a product of being human beings who relate to other human beings within cultures. In any "observation," we necessarily draw on these preunderstandings to structure what we see. (Mascolo, 2016, para. 9)

There is agency in the individual, but it is informed and restricted by context. We can take action but our actions are within the set of actions we can actually perceive. I am careful not to attribute agency where it is not. There is a paradox of agency: I am me, but I am not me; I am also my context.

Change-Making

We are on thin ice here. Are we making change? Or is change making us? Who are we to make change? How do we know the change we are convinced of is "good"? Ultimately, in complex systems, change makes itself. Through research, I have learned that systems do not change; they learn. The term reminds me of the puppeteers who manipulate marionettes, or heroes who ride in on white horses, or patronizing "do-gooders" dipped in righteousness and sanctimony. I am uncomfortable with the distance and hubris the term suggests . . . and yes, I have used it, and in doing so, I only meant that I want to contribute my life to a better world. Still, I am cautious.

Collective Intelligence

What are the differences are between collective intelligence and complicit corruption. Implicit shared understandings, stain and steer our collective sense-making. For example, the idea that material profit is the right of all enterprises. The collective intelligence can easily produce horrible things like destruction of the natural world, racism, fascism, justification for exploitation, etc. I have no replacement term . . . I am more interested in what kind of learning can occur in the liminal realm. However, there is also a sense in which collective intelligence contains the fruit of the collective learning in symmathesy.

Consciousness

Everybody seems to mean something different with this word. I recommend not using this term as a filler for describing culture, epistemology, or mindset. It is an easy go-to. Too easy. The muddle it creates with concepts of Mind, cognition, and that which is not conscious seems to turn everything New Agey. At least for me, this is an unproductive mix-up.

DELIVERABLE
A linear outcome, a harkening of metrics, and impossible to define within projects that engage complex systems change. Forget it . . . after decades of trying to define the "impact" of work in systems change, it is time, I would argue, to give up on deliverables . . . deliver questions.

DEVELOPMENT
I am careful of the suggested time linearity in this term and its tie to progress. Development of economies and structures, as well as the development of people are fraught with ideas of "bettering." Usually, development is a term that refers to an individuated and decontextualized entity. The habit of defining a particular sort of development to a particular entity is likely to negate the contextual information into which the entity is changing. Development according to whom? By what measurement? And with what contextual consequences? It is possible to say that an acorn develops into an oak tree, but that growth is taking place through thousands of organisms. The birds, the soil bacteria, the insects, grasses . . . all are in mutual ecological vitality.

ENERGY
Energy is a word from physics. Be careful. Be specific if you are talking about motivation, enthusiasm, feeling, sentiment, or aesthetics. I suggest using words with relational meaning to replace words from mechanism or physics.

ETHICS
Again, everyone means something different with this term. I find it helpful to ask myself, "Whose ethics?" What is the context in which this understanding of ethics makes sense? We are working in a changing world of multiple cultural perspectives. The old logic of old systems held a body of ethics that may provide a contrast to new insights. I try never to assume that I have understood the ethics of a behavior. However, I like to stay curious and seek more contextual information.

FUNCTION
Pickup trucks function. Living systems respond, anticipate, and learn. Machine language confuses our understanding of more complex forms of responsiveness that are unpredictable in interrelational systems. I use the term "vitality" instead.

IMPACT
The word impact comes from physics. It is what happens when you kick a ball. The velocity of the kick, and the vector and distance of the movement can be measured. Instead, imagine kicking a friend, the repercussions are relational. You can measure the distance your friend stumbles, and the impact of their weight hitting the ground, but how will your kick change your relationship? How will your position in the community be changed? How will your children think of you? When "impact" is applied to living systems, unexpected relational consequences abound. Culture is relational, and thinking about making an impact on culture can blur the possibility of taking into account the interrelational consequences. Even if what we mean is make a difference—be cautious as it carries ghosts that confuse our understanding of systemic change. Social, cultural, and other living systems do not respond to impact like billiard balls. I suggest using the word(s) "influence" or "create conditions for new interactions."

Combining

INCENTIVE
The reason to throw ourselves at the challenge of moving away from our current suicidal systems of living is not to feel important or to get rich and famous. Incentivizing systems change is not a replacement for an epistemological shift toward cultivating better relationships with each other and the environment. In fact, incentivizing is a prolongation of obsolete systems of reward that feed the existing problems. After several decades of failure, the hope of incrementally creating societal change within institutions and organizations has been widely abandoned.

Remember: ecologies do not need incentives. Parents do not feed their children out of an incentive; They feed them out of love and intergenerational survival. Incentives interrupt the real motivations for making change. I try not to underestimate people and thus prefer to appeal to their deeper instincts of love and survival. Anyone asking, "What is in it for me?"—is not yet understanding that the cost of finding a new way of living is money spent. If we are lucky, the return is that there is a possibility of survival. Harsh? Perhaps, but I believe that now is an appropriate moment to take off the gloves. I do not believe that a future of survival is one of material gain. Any future gains will be relational, not material.

INTERCONNECTED
This is a perfectly good word and has brought us a long way. It is, however, better suited for a description of Legos or dots connected by lines than for living systems. In living systems, the connections are relational in ways that are constantly re-fitting. I use interdependent, interrelational, interlocked, entangled, interacting, conjoined, or symmathesy.

METRICS
To measure something requires decontextualization. There are too many variables moving in response to too many variables in the interrelational processes of complexity. Forget metrics; they simply do not correspond to dispersed consequential calibrations of systems change. Metrics are fine for some things, but not this. This process is outside of causalities that are obtainable in numeric forms. Just no. Contextual information is necessary to study living things.

PROTOTYPE
The details of any living system carry contextual patterns. Local zoom-in sensitivity quickly reveals that prototyping across systems, geographies, and cultures is a form of colonialism. I avoid using this idea as it is a dangerous habit to think we can use the same patterns of action in multiple cultures. There are ghosts in this concept. Prototypes are often "rolled out" or "launched," furthering the mechanical context of the word. When multiple contextual interactions are studied, the prototype fails. The question of where it is possible to draw hard lines and where there must be flexibility in systemic understanding is age-old. Alfred North Whitehead warned of the errors inherent in assigning "misplaced concreteness" (Whitehead, 15, p. 64). I hold that advice close at all times.

PARTS & WHOLES
Most parts are also wholes. To be fair, it is very difficult to express the relationships between smaller actors in a systemic process and larger ones without using the terms "parts & wholes." I slip all the time

and refer to "parts" of a system. But, I still believe it is necessary and important to stay alert around this descriptive language of systems. Why? Because the tendency is to imagine parts that are functioning to form a whole. That habituated idea triggers mechanistic metaphors and suggests that change can be made by tweaking parts. What we know is that systems produce mutual simultaneous responses within the interrelationships across and between contexts. I try to stay attentive to my use of these terms and how they tag into ideas of design, arrangement, and engineering, all of which give false insights. I created the term "vitae" to better express the idea of living aspects that we might have called "parts"—but using parts & wholes terminology is a difficult habit to break. I have not succeeded.

Pattern

Patterns change, particularly in living systems, so I refer to fluid patterning. When patterns are sought as such, they are presumed to be repeating static forms that one might be able to identify. This requires a particular form of seeking that is quite different from the way one might find something in motion. Life is in constant fluid patterning. The movement is important, as it shows vitality. If you seek something that is not changing, you will find and report something that does not include life. The fluid patterning in a family allows the relationships to change, the children to grow, the parents to go through their own changes, and the elders to align their attentions in new ways as they move toward the end of life.

Purpose

A critically important aspect of systems work is the recognition that change in a system will change the way the system changes. For that reason, purpose is something to hold lightly. No matter how acute, crises in complex systems are emergent and, therefore, are spinning from multiple contexts, constantly moving into new relational conditions. Sometimes, the idea of purpose is necessary as a starting point, but in systems work, that purpose is bound to transform. Only tiny bits of the larger systems of life across time are visible from a given perspective. Therefore, the benevolence of the purpose of a project is easily misjudged.

The projection of purpose places an end result on a process better understood when open-ended. This is difficult in a culture of silos, goals, and outcomes. Evidenced efficiency toward purpose is likely an expected aspect of discourse around a project; it is a red flag that the project cannot hold necessary uncertainty or ambiguity toward systems change. If the purpose is broadly stated as "systems change," that might be an improvement. Still, with time, that will be a constrictive purpose.

Scale

Scale is a very popular, messy term now and is used to discuss the possibility of identifying patterns of intervention and their applicability to larger systems. It is linked to the urgency of finding global solutions and, like prototyping, part of a vocabulary that suggests ecosystemic, colonial projects. I am acutely aware that the idea of scaling is not an accurate way to describe the relationship between smaller and larger systems because as we look at larger and larger systems, the additional contexts bring in other realms of interaction. Even though the biosphere has variables of interrelation in a similar pattern to my own body, there are very different relational contexts of interaction that are absolutely vital to consider. I am extremely careful with this term. Why not just discuss bigger systems, or zoom out?

COMBINING

SOLUTION
A solution is a linear endpoint to a problem defined in linear causality. Complex problems do not operate within this linearity. Solutions are not what we are looking for. Solutionism is not appropriate in the complexity/systems world . . . If you look for solutions, you won't find them because the problem identified as needing to be solved is a consequence of multiple contextual interactions. There is no hack, no five bullet points, no branded systemic graph that maps the solutions. The thinking that has produced an understanding of our world as a mechanistic, separated, or siloed system also defines our current global crises as such. Food systems, sexism, climate change, addiction, racism, economic inequality, and so on are all consequences of several hundred years of exploitation.

Instead, let's talk about changing the conditions within which interrelational interdependence can take place. Most problems today were once solutions. Proposals that promise solutions to systemic issues are a kind of betrayal to others working in systemic ways who are attempting to stay honest about the benefits of this work. Obviously, the person whose proposal offers solutions is more likely to win the grant or contract than someone whose proposal offers ambiguity. I stand in loyal admiration and solidarity with the latter. The world needs to get used to systemic work having outcomes that are not visible, measurable, or guaranteed. I am pushing hard on this, and I think it is high time we all did. As an alternative, I use the idea of improvement in "responding" to a systemic or complex problem instead of "solving" it.

STEWARDSHIP
The idea of caring for the earth is important. There is a need to tend to the damage. But, as I see it, the imperative project for humanity is to allow for a new way of relating to the earth. The way that relationship is framed makes a difference to me. And when I think of what sort of relationship I would want to be in, the idea of having a partner who wanted to "steward" me is quite unappealing. There is a hold-over of ownership in the term . . . and strange dominance. I am interested in tending to the intimacy with which every day brings the environment into my body/mind/emotion and my body/mind/emotions into the environment. A definition of stewardship is: "Stewardship (n.) The job of supervising or taking care of something, such as an organization or property" (Merriam-Webster, n.d.).

STRATEGY
Again, a linear word. Is it possible to take action and to think in long-term ways without falling into the traps of linear goals? To do so requires an understanding of transcontextual interaction and detail.

WE
Which we? We humanity? We in the room? It is all too easy to romanticize collaboration of a notion of humanity perceived from a position that has not experienced dire exploitation. While it seems crucial to counter the poison of individualism with the more inclusive "we" instead of "I," that "we" assumes similar life experiences where there is difference, and as such, obstructs perception of these important perspectives. The affectionate "we" becomes alienating especially to people who have been betrayed by the existing systems over generations. The attempts at rallying solidarity through the use of "we" often produce the exact opposite and leave many people in acute recognition of the speaker's inability to perceive unique and varying life experiences.

Somehow everything necessary to work for now
seems to float like swirls of sunscreen on the
surface of the pool.

The depth is elsewhere.
May we swim deeply and off script.

GLOSSARY

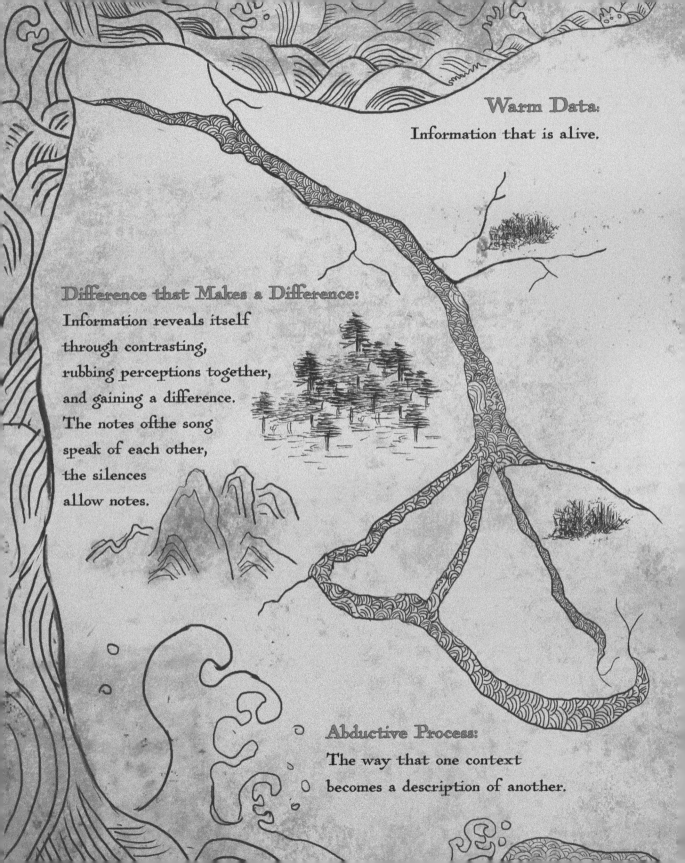

Logical Levels:
Abstractions of the whole—where maps are made by taking out the detail of the territory. The name is not the thing named.

Requisite Variety:
(Ashby's Law)
The multitudes of ways "life is making life" requires multitudes of life-making life. It takes complexity to perceive complexity.

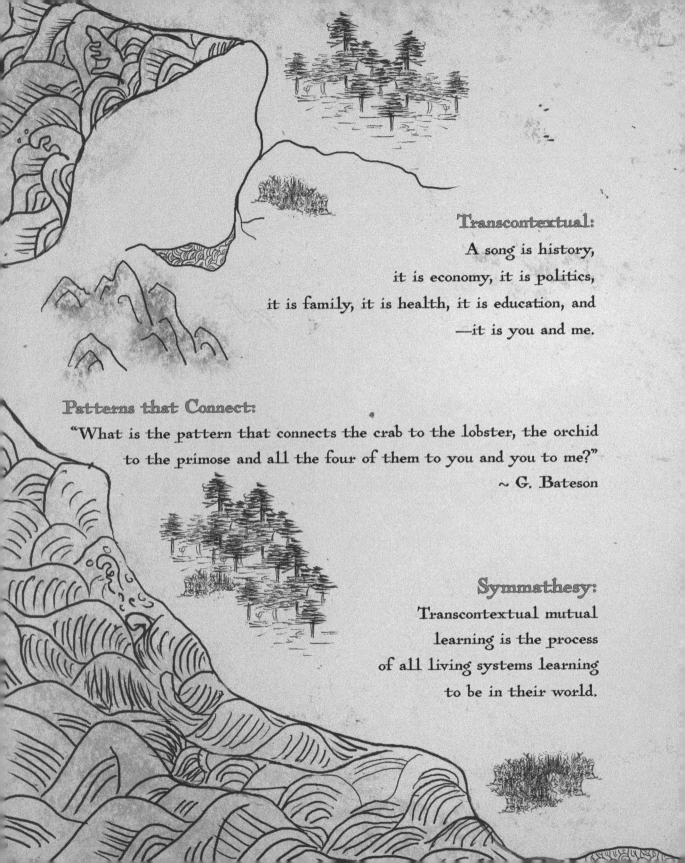

Transcontextual:
A song is history,
it is economy, it is politics,
it is family, it is health, it is education, and
—it is you and me.

Patterns that Connect:
"What is the pattern that connects the crab to the lobster, the orchid
to the primose and all the four of them to you and you to me?"
~ G. Bateson

Symmathesy:
Transcontextual mutual
learning is the process
of all living systems learning
to be in their world.

Integrity:
The way one meets an unfamiliar situation.

Tautology:
The shutting out of all new information through loops that retain consistency by proving themselves. It is what it is . . . but also self-referentiality. A key aspect of both what makes life "life" as well as how obsolescence takes hold.

Multiple Description:
Life is many descriptions—a song, a forest, you, me . . . and the way those descriptions are in relationship to each other.

Epistemological Frames:
A vocabulary of experience,
filters of perception,
the perspectives that are
informed by resonance, bias, and
particular contextual shapings.

Ecology of Communication:
Communication in ecology is more than signal. It resonates into many relationships of both futures and histories ongoingly.

Double Bind: Transcontextual trap, where it appears that failure is the only possibility and there is no way to communicate the experience.

Aphanipoiesis:
There is an unseen coalescence of experiences that is submerging into an un-named, un-known cache—forming what will later be emergence—sometimes insidious . . . sometimes life-giving. . . .

Combining

Ideas are their stories
They are not naked

Ideas are clothed in the experience and languaging of those that give voice to them. Ideas are made of relationships, they travel on and through relationships. They are not smooth bullets flying through the air, they are encounters aggregating between us, the land, our art, and other ideas. . . .

To separate ideas from the relationships in which you or I find them is to strip them of their contexts.

Ideas are not just dangling—they are living in particular spices—brewed in their own sauce of our hurts, our glee, our unique searching. . . .

This makes a big difference that makes a big difference.

We can say that it is impossible to trace the origin of an idea, and that is no doubt true.

But you know where you met, and that meeting is something not to be erased. The meaning of the idea to you is shared in the way you share it.

> The Marxians would, as I understand it, argue that there is bound to be a weakest link, that under appropriate social forces or tensions, some individual will be the first to start the trend, and that it does not matter who.
>
> But, of course, it *does* matter who starts the trend. If it had been Wallace instead of Darwin, we would have had a very different theory of evolution today. The whole cybernetics movement might have occurred 100 years earlier as a result of Wallace's comparison between the steam engine with a governor and the process of natural selection. Or perhaps the big theoretical step might have occurred in France and evolved from the ideas of Claude Bernard who, in the late nineteenth century, discovered what later came to be called the *homeostasis* of the body. He observed that the *milieu interne*—the internal environment—was balanced, or self-correcting.
>
> It is, I claim, nonsense to say that it does not matter which individual man acted as the nucleus for the change. *It is precisely this that makes history unpredictable into the future.* The Marxian error is a simple blunder in logical typing, a confusion of individual with class. (G. Bateson, 2002, p. 40)

Theory is Beautiful

Theory is beautiful. But it is not enough. Until the theory lives in your breakfast, in your walk, in your spine, in your kiss, in your memories, in your laugh, and your tears . . . it will remain disintegrated. This is the moment—every moment is the moment to practice—so that the depth of the understanding is steeped into the fibers of being, and not just floating at the top of a lovely intellectual exercise.

My dad used to say, "It doesn't count until you have it in your elbows." Nice to know Korzybski said essentially the same thing.

Korzybski's warning:

It's not enough just to READ Science and Sanity, even two or three times.

> The reader must translate . . . [the verbal formulations] . . . into his psycho-logical feelings and attitudes . . . [at the object level] "Otherwise he will inevitably miss the point."

For instance, if it is said that,

> ". . . the objective level is unspeakable" the reader should try to become entirely "emotionally" impassive, outwardly and inwardly silent about an object, or a feeling, as whatever we may say is not 'it.' This make(s) "him an 'impartial observer.'" (1994, pp. 328-329)

THE REASONS

The reasons
not to break relationships of vitality are not
reasons, but a forest floor,
In the bone,
In the air-you-breathe-basis.

An assumption, unspoken,
ultimately, infinitely, intimately
steeping,
In the vast vibrancy of life.

The question, "What's in it for me?"
is thus mocked.

SALT AND IRON

The glamor of the modern world is peeling off.

Chips in the paint reveal untold histories
That were there all along.

Soaked
In
Blood

(even the script of interconnectedness feels
sold out)

Older generations will have to say they once
believed in the market.

Told their kids to "succeed"

I remember posh houses
With umbrella holders
Made of elephants' feet.

The whole world is made of elephants' feet.
Shiny shame
Once beautiful and wild

Signals of showy gore.
Enough
Now.

Combining

Fig. 38. Rachel Hentsch. (2023) *To Live in Another Way*. [digital art].

THE ZOMBIE CATERPILLAR

There once was the most adorable furry caterpillar. It had lots of wondrous little legs that moved like rows of dancers as it scooched up and down the stems of flowers.

Each day, it looked forward to munching leaves. Then one day, it heard the news that it would soon cocoon, and it would no longer be a caterpillar.

It mistrusted the idea of flying around. It was unsure of what or who it would be in another form. It became greedy for more time, being furry and many-legged.

It knew life as a caterpillar was a proper and good life, and could not yet understand what it would be like to not be a caterpillar. So, it began to bargain with transformation.

It wanted to preserve itself unchanged. Stealing time and wagering against the meadow it lived in, it hedged its bets and distorted its appetites.

Whenever it felt the urge to cocoon, it made adjustments to sustain its caterpillar-ness until it became a zombie, an un-transformed mockery of the lovely little thing it once was.

The butterfly that it might have become was postponed. The owl eye spots on its wings were interrupted.

COMBINING

Intergenerational learning
is where the change is.

My daughter correctly assessed
that the waste of resources in wars
for resources isn't good math.

Money is a bad investment.

BACTERIA

Are the bacteria in and on your body assisting
you in your immune and digestive systems?
Or is your body assisting them?

Where is the edge of an organism?

Meanwhile, life is mutually vitalizing.

While everyone seems to be attacking everyone for
every imaginable idea.

The shredding of ideas must not shred the deeper
human warmth.

Things are getting really weird. They will get
weirder.

Walk barefoot into the forest.
Drink water from your cupped palms.
Wrap your arms around a crying child.

To get there is not safe, but to not get there
is deadly.

Combining

```
NOCTURNAL

Smearing insight unevenly -
The scratches in the sky contrast
permanence,
Interrupting the notion of time,
So it's possible to find
Another rhythm.
```

BUILDING AN ARC

I dropped it,
Clear glass in water on the floor,
Imperceptible,
Pieces in the puddle.

Broken sharp edges
Of relationships,
That crashed into each other,
Everywhere.

In a time that is cracking,
Everything.
Full hearts spilling,
Everywhere.
With nothing to contain them,

Sloshing into each other,
From between the shards,
Of what once held them,

Safe now that nothing is safe,
To be streams headed down the same hill,
- Finding a way.

While being shaped,
In new formlessness.
Together
Everywhere.
Everywhere.

COMBINING

> Perpetuating existing systems
> that perpetuate existing systems
> that perpetuate existing systems ...
> Of not taking care of life.

Freak Out and Freak In

Things are getting messier. And the messier they get, the more dramatic responses become. Things are getting more peculiar, and they are evoking more unacceptable, uncanny, uncool behavior. So it should be as the stitchery comes loose.

Folks are out there banging on about the need for imagination and creativity... but shunning people who are not acting in a familiar coherent code of communication. Perhaps the ones that are not able to fit into the rules are not the ones to worry about? Perhaps they are the scouts? Perhaps holes need to be torn in the status quo? Perhaps it is a social duty to re-orient?

"Inappropriate" is a word that pops up—and let us ask, inappropriate to what rule set? One has to wonder what it is to fit nicely into a fatal-soul-sucking death box of obsolete rules—and also how is courtesy both needed and suffocating? I remember when my ten-year-old son got in trouble for drawing penises in class. "This was not appropriate." they said. I had to disagree. I am pretty sure penises are an appropriate curiosity of some ten-year-old boys. I explained to him this was not a celebrated art form in fifth grade. It is hard to be appropriate when you are curious and a little bored.

People, including myself, friends, family, colleagues, strangers, public figures, unknown faraway people, intimate, and up close people... all of us are contorting in the various pinches and punches of this era.

History has made us all crooked trees. People are acting weird. They talk too much, they lie, they are obsessed with irrational ideas, they are high on hubris, and writhing in inner despair. It is not easy to be together now that we need each other most. People are getting intolerable.

It comes out as off, offensive, and unloveable. Behavior like this, we are told, one should "set boundaries" against. But this is a time of difficult and confused un-framing.

It is necessary to freak out and freak in—to unravel existing cultures and systems that have framed identity.

This is the rotting time, the composting of old obligations, patterns, and ideas. Getting out is freaking out. I believe that the cadence of the future, the tone, the temperament, the atmosphere... is set right now in how we are with each other as we get funky and disorienting.

Finding unearned warmth to offer in the wreckage. For who among us is feeling either stable or centered in our interactions?

Combining

You never know where "development" is brewing
in the many contexts of our ongoing becoming.

That non-knowing is a deep respect for other
people's complexity.

THE RUBRIC

The rubric will seduce its creator
into believing it takes account
of what is (really) there.

The rubric does not.

The rubric explicitly excludes the
unfamiliar and unnamed moments –

The ones there but not seen through the limits
of the rubric's creators' assumed criteria.

The coming together of time, the happenings,
and transcontextual relationship communing
will always undo the rubric – unless that
evidence is seen through the same rubric.

The rubric will not reveal its bias.

Combining

I don't try to fit life-ing into any models.

I just let the models be distant dreams that cannot be spoken into a morning's kiss.

Lurking Monster

There is a lurking monster whose familiar diet is measurement and control. It lives in language and wears costumes of assumption. The monster twists perception to match and confirm its rightness—unquestioned. The categories it craves are firmly delineated, confirmed as real, describing a non-world of non-living static-ness. The monster loves to compare measurements, "This is better than that." It crunches numbers but defies their poetic beauty. Numbers are stories that shift and slip through each other. But the monster uses them otherwise . . . holds them captive in metric cruelty . . . biometrics, psychometrics.

Divide and devour.

Like Damascus steel, the confirmation of this habit is folded in on itself from fifty directions. (Count the fifty folds if you must, list them, make a spreadsheet for them . . . but you will never control them.) It is the stuff of economy, the basis of psychology, the shape of education, the law of the law, the mayhem of the health industry—and the soul of digital tech.

The lurking monster will funnel the world into the proper slots of the forms. Tick the boxes department by department . . . this will prohibit any healing, any change toward an ecological understanding of the world. It will seduce the process of saving communities and forests into authoritarianism . . . telling people how to live, think, and feel.

Feed it open-ended processes and blurred edges—unfold its thick-tongued speech of reductive confinements and see what it does when it gets dizzy and lost.

There—right there . . . in the vertigo . . . the costumes of assumption are visible as such, the strings of the puppet are in sight, and the monster's influence becomes dusty . . . obsolete.

There is no need to fight this terrible thing; fighting it makes it stronger—better just to notice when it creeps into the moments of the day.

When you see the dog poop . . . you don't step in it. You just don't.

This is why I do what I do.

WHAT IS SANITY?

There was once a student who asked my father, "What is sanity?" My father hemmed and grumbled in disapproval of the question.

After a few moments, he said, "I suppose sanity is familiarity with your own epistemology."

Something to think about.

What is familiarity with your own epistemology in a world in which the structures that define and orient identity are pixelating?

Common Sense Is Sense-making in the Commons

People often said my Nana had common sense: she lived to be over a hundred. When my family uses that expression, they generally mean she had a grounded way about her, not easily taken in by insubstantial notions or flashy distractions. She was who she was and did not go for the conspicuousness of boasting or puffing up. As such, she could spot dishonesty in a heartbeat; she was a no-nonsense woman who knew what was important in life and what wasn't.

That is one form of common sense–the common sense of being practical and of sound judgment. But perhaps this common sense is informed through another where both the "commons" and the sense-making come together so that the premises and basis of communication are mutually produced between people.

What if one understood common sense as a mutual co-understanding? Standing under the other things we do, there is a ground—a common ground of assumptions where the most pernicious of limitations exist. The ground is there in the grammar of the descriptions of the most profound and mundane experiences of being human, in the codes and sensibilities of daily life and growing up, placing ourselves amongst each other in the social landscape. These codes differ between peoples, families, and geographies from one era to another—but they are always there. The common sense is what glues the cues into how to read a circumstance of safety, danger, love, jealousy, inclusion, and exclusion. The commons of our sensitivities is an ongoing, ever-shifting relationshipping with each other and the world around us. Contextual recognition and confirmation are communicated through the expression of (or the lack of thereof) things like courtesies, endearments, terms of respect (and disrespect), as well as the signals of wealth, position, and friendship. Until there is the experience of being in a set of common sense codes that are utterly unfamiliar, these mutual understandings amongst a group of people may be entirely unnoticed and even assumed to be "human nature."

The sense shared between us—the unspoken logic of mutual responses, the familiar and understandable pathways of how to get from today to tomorrow—is where a sense of the unsaid commons of our communication is made. The unsaid and unnoticed presuppositions are far trickier to point to. These common assumptions are in the air and water; they are laced invisibly into language. The sense we make together is not my fault or yours, neither my doing nor yours; it is not point-to-able or easy to pluck and reshape. The commons of our sense-making is a realm where the way we learn to be our world together is written through our relationships, formed and informed into our days with seamless becoming. The ideas in common sense are woven into everything: health, education, religion, science, history, politics, culture, economy... identity!

I had the honor of being invited into a tea house with a friend in Kyoto, where I found myself in a world that I could feel was rife with ancient, refined, particular codes of communication—none of which I knew. I was unsure where to put my hands or whether to smile. I did not know what to talk about. Was family a good topic, or perhaps it was not okay to discuss personal life? Should I eat the fish before the rice or the pickle? Should I bow? If so, how... lightly, deeply, only with eyes? I was entirely out of my world of codes and in another. I realized I had no idea what to do nor how to sit or hold myself. I probably botched the whole thing terribly. In that experience, I was oddly reminded of how many millions of codes of behavior I know—although in other circumstances. I had never noticed the enormous bank of cues and scripts I was carrying until I was suddenly someplace where none of them were helpful—like a language I don't know (and I don't speak Japanese). I have many words in my tongue, but I could not use them to communicate here.

The languages of gesture articulate the details of particular sensitivities—the volume of voice, use of hands, adornment of the body—that implicate larger sets of shared ideas about how life can be lived. At that moment, I glimpsed a tiny peek into how it might be for many people for whom symbolic and representative communication, whether it is verbal or non-verbal, is confusing. I was outside the common sense—outside of the shared assumptions; it was disorienting in a way that reminded me how it feels to be in the familiar of my world in contrast to having no way to know what I was or was not communicating.

It is too easy to say that one should simply "be oneself" in such circumstances—it sounds good and seems at first like there is a "just be human" vein of communication from which we can all source. But I must admit that my "just be me" possibilities were scrambled. The "me" in this context of the tea house was in a new set of relationships, stripped of my history, stripped of my voice, stripped of my codes. I only knew that I did not want to be offensive, that it was important to me to show that, even in my blundering way—I wanted to transmit respectfulness. I hope that I managed. Isn't it incredible how one learns to be in one's world? There are mountains of almost invisible information about how to share a communication zone.

My father was interested in how the relationships and communication between people were a realm of an *Ecology of Mind*, where ideas could grow, seed, die, learn, evolve, wilt, and sometimes become obsolete. The importance here is noticing that these ideas are held among and between us, between and among us, in the air we share, and in the implicit codes of culture.

Like a forest, ocean, meadow, or your body, an ecology of ideas requires many organisms to become, continue, and change. The jungle is not in any individual organism, and the jungle keeps jungle-ing through the ever-shifting responses of millions of organisms responding to millions of organisms.

I cannot undo my sensitizing to individualism alone. I cannot untrap you from seeing me in particular categories that then trap me. This is not an explicit project in which I can be an advocate or activist; I cannot vote against it. It is impossible to disown the common sense. We make it together, and we must unmake it together. What are the underlying premises of how to perceive the world?

The basis?
The is-ness?

What is in the soup of those presuppositions upon which everything, everything, everything is balanced, justified, and continued? And what happens when that commonality is premised in the ideas and metaphors of linear causal control? Of individualism? Of a misunderstanding of human participation in the living world of organisms that together produce vitality? What happens when the common sense is contingent upon an illusion of separability?

If individual-ness is the world perceived and lived within, then that is the world you and I will respond to. But as is inherent in the term, the catch is that common sense is not yours to make. It is made like the air around you; you are in it—we hold one another to it without even noticing we are doing so. It is insidious; it lurks and is very difficult to alter.

It is certainly nice to try to "be the change"—but even that statement, as lovely as it is, infers the causation at the individual level again. Inherently, it reveals that the common sense is centered on the illusion of the individual, which is not how life is. Life is interdependent. While it is easy enough to use this vocabulary, the depth of the implications of interdependency is unfathomable.

What if we are all trapping one another in the limitations of "common sense?"

What about symbiosis? Right down to the infinitesimal expressions of life, we see that the formation of cells is contingent upon many organisms. There is not one organism that is one organism to itself. You are mostly made of other organisms, and the cells identifiable as "human" are produced through long combinings of other organisms. Your senses are produced through pathways of other organisms.

What is sense-making, then?
What is common sense?

My father once said:

> We live in a world in which distrust and greed and violence masquerade as common sense, and in which the pathways of distrust and greed and violence are rapidly becoming self-validating. By following those pathways we create the social and international structures, the premises upon which we must live. By choosing the "common sense" of distrust, we choose also the progressive truth of distrust. We cause horror to become the only pathway to wisdom. (G. Bateson, 1979, p. 36)

It may seem he is pitching for greater amounts of "trust, generosity, and non-violence" here. But that would be missing the deeper ecology and the systemics of what he is saying. He does not imply trust and generosity are better versions of common sense that should be propagated. No. The issue he is pointing to is that mistrust is a 2nd-order response to a world whose common sense is premised upon a way of understanding our world centered on individual, separated, controllable, ownable, measurable bits. Probably, it's easier to completely overhaul the premises than to try and make adjustments one piece at a time. The common sense premises of a world that is perceived to be fragmented will always include notions of separated entities engaging in justifying ideas of:

> **Linear Causality**
> If you are only attending to one separated part of a system, you will not see the ways in which many organisms, ideas, cultures, and time come together to form shifts.
>
> **What is in it for me?**
> If I am separate from you, then my world is mine. If I perceive I am separate from you, I cannot perceive the trillions of organisms I am made of.
>
> **Material profit is a right.**
> Because everyone has to make a profit in a world of individuals, obviously.
>
> **Exploitation**
> Objectification erases the relational jungle, making it possible to justify the continuing exploitation of people, nature, and life.
>
> **Appetite for measurement of decontextualized aspects of the system**
> The blur of relational movement is unseeable due to objectification, and these decontextualized measurements are "posing" as information.
>
> **Appetite for methodologies of control of the parts of a system**
> Control, even as an illusion, is preferable. The blur is messy, and it takes rigor and different habits of perception.

The list could continue; the soup has brewed so much destruction and pain—from mental health to ecological health, the common understandings of a culture that does not understand symbiosis as a formative principle of daily life are caught in its own traps.

> Living beings defy neat definition. They fight, they feed, they dance, they mate, they die. At the base of the creativity of all large familiar forms of life, symbiosis generates novelty. It brings together different life forms, always for a reason. Often, hunger unites the predator with the prey or the mouth with the photosynthetic bacterium or algal victim. Symbiogenesis brings together unlike individuals to

make large, more complex entities. Symbiogenetic life forms are even more unlike their unlikely "parents." "Individuals" permanently merge and regulate their reproduction. They generate new populations that become multiunit symbiotic new individuals. These become "new individuals" at larger, more inclusive levels of integration. Symbiosis is not a marginal or rare phenomenon. It is natural and common. We abide in a symbiotic world. (Margulis, 1995, p. 14)

What if symbiosis was the basis of common sense, as it is the basis of everything alive? What if symbiosis was the premise upon which humans built the ideas of success, health, learning, risk, home? Is this not practical, no-nonsense, and sound judgment to live within?

FREQUENCY

Smearing insight evenly –
The scratches of the sky
Contrast permanence,
Interrupting the notion of time,
So it's possible to find
Another rhythm.

(THERE IS NO SCRIPT)

It is because I care so deeply that I get grumpy and tired of the scripts and methodologies for self-help, saving the world, and the best way of being "interconnected."

There is no script, plan, or formula.

It's instead about the way healing seems to happen when dignity is implicit: There isn't a single thing that can be said that is not going to be attacked or amplified ... or both.

Maybe this is what it takes.

The scripts won't help. They only signal the boundaries of belonging ... and those boundaries are moving. They have to move.

The scripts are not conversations. The scripts wall the mutual learning.

Everybody knows a script when they smell one. They know they no longer have to pay attention to the conversation. They know how their own script sits in contrast.

The scripts are boring anyway.
Freeform this.
Go on.
Fling your arms in odd gestures.

Minutiae of the Day

We could even convince ourselves that it is not within "human nature" or the capacities of "human cognition" to be able to perceive ecological processes. They are too multiple, always changing, too entangled.

I am not so certain about this.

Ancient cultures and indigenous peoples have participated in ecological processes for thousands of years without destroying them. They knew. They did not use the vocabulary of complexity theory established in universities in the 1950s, but they knew.

I work with small children who grasp these concepts, as well as people all over the world who have never set foot inside a complexity theory class. Life is already doing this. Being able to talk "about" complexity and ecology is not the same thing as being—living with a perception that is within the ecologies and complexities of life.

About-ing is not the same.

And anyway—we are only partially human. Most of our cells belong to the trillions of organisms that live on and in us. Perhaps they are as yet untamed by the industrial dream? Perhaps we have sensitivities that are still within the membership of ecological process—it is not a professional tool. It is a way of life that is all-encompassing and saturates through every minutiae of the day.

Getting "complexity" begins at home in the communication and perception *within* our most intimate relationships.

This is vital.

Fig. 39. Rachel Hentsch. (2023). *Meet Not Match*. [digital art].

IN THE FIRE

Every behavior makes sense within particular
contexts. Be careful what contexts get created.

But now that we are in the fire ...
Responding to the context and not polarizing
with the behavior is enormously challenging.

The horizon is all questions, play, and
tenderness.

And some rage ... some tears.

This is a diffused process.

Tearing and Mending
Transcontextual Learning and "Healing"

> The fabrics of interaction are going to be torn and they're going to mend. And after all, we are only a model of what we are trying to talk about, and it would be absolute nonsense to try to construct that model as though it did not contain the tearings of fabric. (G. Bateson, 1972, p. 293)

Can you hear the sound of fabric ripping, hypocrisies bursting out in little torn spots here and there? Through these raw holes, language is beginning to be woven around the sensitive and complex presuppositions about day-to-day life that must be brought into the realm of our conversations. Established patterns of communication appear to be pixelating, unable to hold the complexity that will no longer be silenced.

News media, academic research, political discourse, market analysis, and legal systems are hamstrung by their own capitalization, losing credibility by association with the very history they were previously accredited. It will not be easy to decontaminate these patterns of communication, nor will it be convenient to re-think the ways in which their mutual influence has brought goodness into the world alongside exploitation and a history of horror. In addressing these conflicts, it becomes clear that the conversations are the stuff of both survival and destruction, running in the same veins, carried in the same blood. Prising them apart is not an option.

Many beautiful minds and hearts are at work now, trying desperately to meet the dangerous era of interlocked crises. With each passing day, the possibility of collectively meeting these crises is undermined by increasing ideological polarities. Ideological divisiveness is turning family members against each other and breaking communities. The divisions spread like cancer, repelling people who might otherwise be ready to help one another in need. The word "trauma" is soaking through almost every conversation, so much so that its meaning is becoming confusing. People are holding profound fragmentation in their beings. The need for people to be able to need people as well as be needed is looming large in the face of upcoming decades of social structures being undone by political, economic, and ecological events.

And yet, seemingly, we cannot talk to each other right now without feeding the divisions. The past has been so full of hypocrisy and justifications that inflammation has become explosive tension. The infection is everywhere in the form of one thousand kinds of abuse—filling classrooms, churches, homes, hospitals, markets, fields, offices, playgrounds, oceans, forests, and factories. We, as humans, need community, but *before community, we need to commune.* Familiar patterns of communication, language, and cultural scripts seem only to make more divisions. It has taken so long. Now, the wounds are septic with a need for speedy recovery, which blasts through the patience of slower learning that the injury can offer. In the singularity of its mission to hastily fix one malady at a time, the cure may be more harmful than the wound. The tearing apart is necessary. The mending is also necessary.

The notion of "healing" carries all sorts of metaphors from a world of industrial systems in education, health, economy, agriculture, media, technology, and so on, as well as the subsequent justification for ongoing destruction that these systems generate. It is a dilemma that "healing" from industrialism will require an entirely new way of thinking about healing. Of course, ancient forms of tending and nourishing have always been more reflective of natural processes.

The metaphors for tending and healing are repeated again and again in the natural way in which the ecology of relationships around any organism has healing within them. However, the intrinsic healing rhythms and relations have been disrupted in this fragmented and decontextualized world. Our approach to an ecological form of mending will be new only in so much as it is inclusive of learnings from the mistakes of the industrial era. I hold this learning about to be crucial now. There is no way to go back. Learning from here must include the scorched blisters of having touched the fire and the deeper burns submerged beneath the skin.

The healing will include the burns, the pus, the fire, the grief and loss, the scar, and the cellular reflex to be more careful—lest it is forgotten that the vitality of living systems will not be mocked. The mistakes of reductionism, the misconception of individualism, and the errors of predetermined linear outcomes of healing all validated a slogan of "better, faster, cheaper." Other realms of crisis are also currently rubbing against the issue of solutions that are perpetuating the problems. The need now is for more nuanced and contextually responsive ways to describe the issues so that other pathways of inquiry aside from linear solutioning move into play and begin to take form.

The people who bring healing into their daily encounters with other people are everywhere. They can be science teachers, activists, permaculture farmers, artisans, software programmers, bus drivers, musicians, social workers, doctors, cooks, mothers, fathers, forests, novelists, community organizers, and even "gurus." Science is not an anvil to cut apart healing from healer—this division between the perceived healer and the perceived patient is a gulf of lost relationship. The characteristics of healing are as infinite as the characters it happens in relationship with. Is the healing in the healer or the relationship?

> Life on earth is more like a verb. It repairs, maintains, re-creates, and outdoes itself. (Margulis & Sagan, 1995, p. 14)

Healing is dispersed; it is not only in the spot that hurts. It takes many contexts shifting to support a change in the identified problem zone—a good friend, long walks, fresh food, music, soft blankets, and ideas that change. Perception changes in small flashes, watery seepings, frantic scribbles, savored flavors, long silences, and funky beats. People worldwide want to contribute to the possibilities of humankind; it is a rush of warmth just to think of it. Even in misconstrued attempts to make positive change, the drive to show up and try is significant. The challenge is how healing can be sucked into the epistemology of fixing (presupposes broken), which is a worshiping of the industrial idea creating the infection. Such are our times.

The Accidental-ness

Five years ago, I accidentally did a Warm Data Lab session with a group of people with severe mental health diagnoses. It was an accident because I was unqualified to be with them in that capacity. There was a botch up of communication, and I thought they were activists, while they thought I was a therapist. I did not have the slightest clue what was going on until we were already well underway. By then, I had given a presentation on the difficulties of parsing out information about health with reductionist research. I prepared the group for a Warm Data Lab, as I had done at many conferences around the world. I had no idea. They had no idea. It was a perfect mistake. The organizers were as confused as I was, and to this day, I have no idea how this happened.

After an hour or so in the process, one of the participants (which is how I saw them, not as patients) said something like, "This is the first time anyone has come to meet us where we are. We can see now that we got lost in the warm data, in the misrelating of relational perceptions—in this process, we are finally given a place to allow those perceptions to find new relational connections." With tears in my eyes, I wondered how one could even respond to something so profound. What can I do to offer this to others? How do I hold this? What the hell just happened? Because of the accident, a peek into a realm of possibility opened.

However, I did not leap on it, scale it up, and put it on the market. I did nothing of the sort. I was hesitant to explain what happened for fear of sounding like I was exaggerating. Furthermore, if someone believed what had happened, there would be no way the existing forms of therapy could make sense of the causality of how this had happened. In fact, I could not even imagine how to approach the research to follow up on the group's insights, let alone how to get that research funded.

But something had happened. I had previously seen a similar phenomenon—a recognition of a release. I had occasionally described it as the feeling when a tourniquet is removed, and the blood flow returns to a limb that was becoming numb. Interestingly, that release is often surprising to people who have not noticed the constriction. Once the participants begin to perceive (in a felt sensing) the release of these limits through the transcontextual process, they begin to experience their own memories of school, for example, as being not just about school but also about culture, history, economics, family, and identity.

Before the "accident," I had only witnessed this release in circumstances when the people participating were not collectively identified as "mental health patients." I made a few feeble attempts to get some research going, but I was not fierce enough in the end. In time, I would understand why I did not pursue that avenue of research. I was waiting, listening, and paying attention to what I was witnessing. I was trying to find language for it. Trying to taste, hear, smell, and texturize the difference in the quality of the shifting I was perceiving. I made new words. I sat quietly in them. I waited and participated in hundreds of Warm Data Labs with thousands of people. For years and years, I have wrestled with the language of how to describe that day. I have wondered whether I should describe it at all. I have worried that if I spoke of what happened, people would be eager to "use" Warm Data Labs as an instrument, a tool, a therapy . . . precisely what I do not want to happen.

This is a predicament that has generated a significant struggle. After hundreds of Warm Data Lab sessions, I can wholeheartedly say that something happens in the group akin to suddenly taking a breath after being underwater, but in the form of a cognitive, emotional, and psychological belly breath. Part of what allows that to happen is that the sessions are always without agenda. There is never a collecting of insights or a harvesting of comments. Whatever opens will find a way—making it a line item on the action list is a flattener. The impetus that bubbles up must not be harnessed. Like with art—the combining and experience are beginning to cook in the depths of each participant and find expression wherever the contexts resonate. Interestingly, curating such combinations resonates with "cure," curating is about selecting a collection of experiences to let meaning emerge.

I have been shy to discuss my experiences with Warm Data Labs. I am reluctant to expose this beautiful, spontaneous depth of experience to the possibility of being yanked from the rich soil where it is being nourished. The small step between the Warm Data Lab being a slightly mysterious process that pops up in conferences and other community events to it being a treatment tool is likely to get momentum in all the ways in which the processes have been carefully crafted not to. I do not want the Warm Data Lab to be used as a medicine or remedy for a particular purpose to solve a problem. At that moment, the whole thing would lose its underworld of rich connective processes that bring ripples of change at nth-order.

A challenging and terrifying irony is inherent in the project of letting go of those promises nested in industrial thinking that offer relief from the suffering generated by the same thinking. The healing becomes engineering if the living world is perceived in mechanistic metaphors. If the living world is likened to a machine, then the question is, "How do we fix the parts?". . . naming the problem, making a methodology to solve it, and measuring the outcome. This predicament has been an invitation into a world of difficult questions. What would something look like that was "healing," but was never identified as healing? If people wanted to participate in Warm Data Labs because they were *told* it was healing, how would that change the experience? How do we take care of the ember of possibility so that it remains unique to each participant in each context? Does identifying Warm Data Labs as therapeutic push healing and relief into a linear outcome? Will people come to a Warm Data Lab for healing? And what if they don't find it?

The Warm Data Labs are not about solving identified problems. Instead, they allow movement through many aspects of memory and perception that can alter the underlying assumptions about who I am, who you are, and what life is about. There is no way to know what will change, when, or with whom— which is fundamental to the theory and the Warm Data Practice. The effect and its cause are nurtured indirectly, never action-ed, named, or strategized. The way changes occur when people are not caught in direct correctives allows for the particular experiences a person had in the past to re-combine in their sense-making—in stochastic, alive, and utterly unpredictable ways. These re-combining movements spread throughout people's lives, and they often later speak of sudden changes in relationships with their children, spouses, parents, and community. But also, surprisingly, people speak of physical shifts, changes in hearing, muscular coordination, and sense of taste.

The process produces overlapping stories, ideas, and perceptions at a high rate and concentration. Impressions light up and are re-framed, re-flavored, re-storied. There is what I refer to as a moiré—a phenomenon of patterns of communication, differences of context combining, and tone among multitudes of other coalescing information. Imagine the experiences of your life like vegetables in bins at the market, school experiences, work experiences, family experiences, the general experience of being in your culture, religion or spiritual experiences, and experiences in nature or with technology—and then think of these stewing together into a soup. For many, this soup is just the soup they live in. The warm data is the combining of these experiences, and the Warm Data Lab allows for the recombining. When they recombine, a new soup of information exists for people, and old memories are seen through new contexts. They continue to recombine . . . sometimes for months after a single session.

As a host of Warm Data Labs, I would never want to manipulate that combining process or declare what I, as a host, believe to be a more desirable soup. For me, this is a matter of crucial integrity to the process. I am not hoping to change mindsets or alter people's lives. Instead, the process lets them meet the details and underpinnings of their world, and from there, the movements begin. It is not therapy. About one is there to talk "about" anything in particular. One never knows what story or what impressions will surface. There is no sharing of accomplishment or competition. Often, people do not think anything has happened at all—aside from a warm conversation—and then, a few days . . . a few weeks . . . a few months later, they realize they are looking at the world quite differently. Maimunah Mosli, who has been practicing with Warm Data Labs in Singapore for several years, remarked:

> Warm Data Lab processes, to me, offer a sacred space that I am still learning within and continue to experience in the many labs I have attended. It offers the space for "no answer." It offers a space for pausing. It allows everyone in the room to ruminate with what matters to them emotionally, psychologically, spiritually, and physically. It piqued for healing to lurk. To me, the permission to be in the space where our thoughts, our emotions, and making our "being" matter is an essential beginning for anyone who has yet to feel that the "self" matters or has never mattered to anyone else.
>
> The process made people feel their stories, their lives, and their "being" matter. This mattering of the self and the context of where one is or what one is deliberating on is the sacred space that many healers, practitioners, helpers have traded—we have traded the "healing process" for mechanistic, management funding modalities. (Personal conversation, 2022)

Combining

If I hold the Warm Data Lab space well, I will never know where, when, or if the "healing" may happen. It is not mine to know. It may be that a participant speaks differently with their children, and the children respond to being in a new communication with a parent—so they treat their friends differently at school the next day. Those friends may then treat their friends or parents differently, and so on. The work of systemic tending is the work of nth-order change. I will never know what happened. I love that.

Most identified problems, as they have emerged, are the consequence or symptoms of other conditions. The solution to the consequence is likely to perpetuate the actual problem. In the previous chapter, "Aphanipoiesis," I discuss this in more detail. [see page 145] The same can be said of larger issues like climate change; the problem may be carbon particles in the air, but the long history of cultural respect being produced through material wealth has guided the behavior of . . . (all positions in the wealth gap) . . . fueling the destructive and exploitative industrial production-distribution-waste cycles. The issues are formed upstream from the emergence in the earlier overlapping and combining of unexpected and subtle experiences. Those subtle day-to-day subterranean learnings form and continue to shape each of us.

> This unseen realm is vital, non-trivial, and sacred—and it is real. I am increasingly finding that the most fecund realms of change, learning, and evolution are beyond the organism's current capacity to perceive. The flexibility that lurks below conscious perception is like the soil beneath the forest, teeming with relational processes. While most attention is caught up in what can be perceived, there is a wildness in the implicit correlations, connections, and coalescing impressions. (N. Bateson, 2021, p. 2)

What Is Pathology?

Let us start with the question, "What is being healed?" The attention of the therapy or treatment is usually directed toward an identified "pathology"—a term that is worthy of scrutiny; perhaps some new approaches to perceiving the so-called "problem" to begin with will alter the response to it. What is pathology? Merriam-Webster offers this definition:

> pathology (n.)
> 1. the study of the essential nature of diseases and especially of the structural and functional changes produced by them
> 2: something abnormal:
> a: the structural and functional deviations from the normal that constitute disease or characterize a particular disease (the pathology of pneumonia)
> b: deviation from propriety or from an assumed normal state of something nonliving or nonmaterial (the pathology of wine)
> c: deviation giving rise to social ills (n.d.)

If the identified "problem" is a recognized symptom or syndrome, the treatment or the healing will also be perceived as the relief of the symptom. The vulgarity of the notion of normalcy, rooted in Gaussian statistics and eugenics, defies the basic understanding of symbiosis. Every organism is an ongoing "becoming" of millions of other organisms. Defining "normal" requires the removal of the history of combined situations and organisms that have co-produced the patterns. The possibility for deviation is enormous when all the other organisms are brought in, and when the additional variable of "situation" is added, these possibilities multiply exponentially. The zone of "normal" also depicts "abnormal" as that which should be fixed (made normal) or excluded from the "functioning" system. Deviation from a system that is destroying life is somehow a pathology, while fitting into the structures that uphold and justify these systems is often considered "healthy."

"Functional" is a word often used in ways better suited to machines. It can imply surviving, but usually, it implies parts that are in operation together to produce a desired outcome. Each component must do its job of maintaining this clockwork, which is fine enough if one is working on engines, but ecologies never do one thing at a time. Every aspect of a living being is in interdependency with many other living beings, and at the same time, they are all responding to various happenings. The deer's antlers are protective weapons to fight other bucks or a mountain lion, they are an attractor for females, and they have a fuzz on them in which a particular fly breeds—when the antler drops, it becomes home and food to many creatures living on the forest floor, their excrement goes into the soil and provides minerals to the bacteria, the trees, fungi, and other organisms are all also in such multi-relational processes.

Together, they are life making more life. The notion of function in this ecology misses the dance and how the dance has evolved over epochs. Furthermore, the notion of health requires a fractal, improvisational mutual communing—much more than functionality.

Within this definition of pathology, there are clues to the implicit ideas or premises of ideas immanent in the context that produced the definition. The definition tells us quite a bit about what is seen as unhealthy and, from there, what sort of responses are logical. We have already noticed that these notions form a complex nest of inter-validating ideas like the deer's antlers. The idea of functionality depends upon parts with roles defined within normal limits, and this normality is fastened to the idea of deviation and abnormality. Jonathan Goldsmith, a therapist, and Warm Data Host, asked:

> Where did we learn that suffering was something to be pathologized, something to be worked with, to be "treated" in therapy as something that can be fixed or resolved? Our clients often come to us asking us to remove their suffering, to show us the specific tool they can use to remove their pain as if it is a part of the system that can be simply cauterized and rewired. Instead, we can choose to explore with our clients how suffering, when it appears, can become a vital part of their existence; it teaches how to "be" in the world, and held in relationship, it fosters the possibility of change, growth and evolution. We cannot and should not separate suffering from our humanity; it is the sharp tang of bitterness that allows the sweet flavors to permeate through. (Personal Conversation, 2022)

There is a collective subscription to a set of premises that interlock the various institutions into a double bind. For example, one can only be a teacher with a teaching credential from within the system perpetuating reductionist education. It is often not even legal to build architecture in a way that would be genuinely ecological. The permits for various materials do not even exist. These double binds reach insidiously into every aspect of our lives, including the most intimate notions of identity, family, and food. The premises as "rules" of a collective are not necessarily explicit. Still, they lurk under each moment of the day, salting each communication with accepted and expected motivations that are difficult to grab hold of because they are communal. It is a tautology caught in a retraumatizing eddy.

Meanwhile, before the turn of the century, the mid to late 1800s marked the arrival of the factory, bringing another way of perceiving that caught on and began to link reductionism, productivity, economics, mathematics, racism, and exploitation. Eugenics promised higher productivity, control, and ways to have agency that could be measured. These ideas are still steeping into psychology, education, health, politics, technology, and environmentalism. The use of the top-down model to manage and control the way people live, work, eat, feel, and think is often coated in altruism of making them healthier, but this approach is in itself violently unhealthy and pathologizing.

> Ideally, statisticians would like to divorce these tools from the lives and times of the people who created them. It would be convenient if statistics existed outside of history, but that's not the case. Statistics, as a lens through which scientists investigate real-world questions, has always been smudged by the fingerprints of the people holding the lens. Statistical thinking and eugenicist thinking are, in fact, deeply intertwined, and many of the theoretical problems with methods like significance testing—first developed to identify racial differences—are remnants of their original purpose, to support eugenics. (Clayton, 2020, para. 4)

The description of the cure and the disease are both located within an epistemology that has always been dangerously fueled by a need to control, label, flatten, and reify.

Recently, a dear friend mentioned that he "took a leave of absence to take care of his mother." What does that sentence tell us about the world in which one must become absent to give care? Intergenerational care is contraindicated to providing for the needs of the family! Something has gone very wrong. That wrongness runs through the entire way of life—validated at every turn. Elders do not want to be burdensome, and teens are taught to feel they are "failures to launch" if they do not individuate at the appropriate age.

Meanwhile, changemakers are calling for localism, community, and collaboration. The truth is, this stuff is challenging; many people cannot stand to live with their elder parents, and the ideological train wreck is just too much. Nor can they stand their teenagers. Perhaps the pathology is in the idea of individuating at all? Or in the communication that holds these individualistic mandates generation after generation? This is especially the case in North America and some European countries, but some parts of the world have not lost intergenerational living as a way of life.

The elder care homes and daycare centers, university dormitories, and summer camps all allow for the workforce of adults to continue being inaccessible to those relatives who most need to be cared for at home. Often, wealth is not worth the loss of the possibility to care for family. This loss becomes a source of discontentment, depression, and meaninglessness. It is no wonder that for many, home is not always a haven of love. For some, home can be a nightmare. In which case, where is the community? If the symptom is cured, the conditions that generated it are likely kept in play. Go back to work. Get yourself back in the saddle. Get well soon cards could well read "get normal soon"—fit back in soon. "So the doctor who concentrates upon the symptoms runs the risk of protecting or fostering the pathology of which the symptoms are parts" (G. Bateson, 1979:1992, p. 296).

In one direction, there is the cathartic and vital increase in attention to upturning colonial epistemologies; the horrors of centuries of history tangled up in industrial metaphors that have consumed generations are finally coming out in new ways. In another direction, there is the hope that the specialists, trained within the same epistemology and metaphors, might be able to offer some medicine, some relief, recovery, healing, therapy, or treatment for any or all of the many devastating consequences of this history including everything from family abuse to industrial abuse to substance abuse to the exploitation of the water and soil, and the many forms of racist, sexist, justifications for putting the comfort of some people above the survival of others. It includes the many trillions of organisms that are subjected to being used, overlooked, and sold. The irony is that the pills and practices that will make it possible to get up tomorrow morning keep us within "normal" levels of health, allowing ongoing participation in the same systems generating the physical, mental, emotional, ecological, and inter-generational pain. And yet, tomorrow is always on its way.

The idea of healing is caught in a linear outcome-based attention to symptoms. Treatment as a concept is again snagged into the notion of an implementable strategy toward a predetermined description of the success of the treatment. Healing is for sale; it is sped up, optimized, and made into a war of patients fighting their diseases—winning and losing battles against an enemy. Even though separation from the pain is the "goal," the whole ecology of the being in all its relationships will bring its stories with it. The learning is located in how to travel together, find out what works, and allow change to unfurl as needed. Healing does not end, and it eschews any predetermined arc. It can be parasitic, symbiotic, or both. In any case, it joins the cast of characters in the story of a lifetime.

"We are all just walking each other home." ~ Ram Dass

Frankenstein-ing

There is a story my father used to tell to illustrate the hazards of cherry-picking what "preferred" changes in a living organism might be and manipulating them into being. It is the sad story of the polyploid horse, optimized to be everything more of what one man thought was desirable in a horse, and the poor animal became a kind of Frankenstein being made more productive:

> They say the Nobel people are still embarrassed when anybody mentions polyploid horses. Anyhow, Dr. P.U. Posif, the great Erewhonian geneticist, got his prize in the late 1980s for jiggling with the DNA of the common cart horse

(*Equus caballus*). It was said that he made a great contribution to the then new science of transportology. At any rate, he got his prize for creating—no other word would be good enough for a piece of applied science so nearly usurping the role of deity—creating, I say, a horse precisely twice the size of the ordinary Clydesdale. It was twice as long, twice as high, and twice as thick. It was a polyploid, with four times the usual number of chromosomes.

P.U. Posif always claimed that there was a time, when this wonderful animal was still a colt, when it was able to stand on its four legs. A wonderful sight it must have been! But anyhow, by the time the horse was shown to the public and recorded with all the communicational devices of modern civilization, the horse was not doing any standing. In a word, it was too heavy. It weighed, of course, eight times as much as a normal Clydesdale. For a public showing and for the media, Dr. Posif always insisted on turning off the hoses that were continuously necessary to keep the beast at normal mammalian temperature. But we were always afraid that the innermost parts would begin to cook. After all, the poor beast's skin and dermal fat were twice as thick as normal, and its surface area was only four times that of a normal horse, so it didn't cool properly.

Every morning, the horse had to be raised to its feet with the aid of a small crane and hung in a sort of box on wheels, in which it was suspended on springs, adjusted to take half its weight off its legs. Dr. Posif used to claim that the animal was outstandingly intelligent. It had, of course, eight times as much brain (by weight) as any other horse, but I could never see that it was concerned with any questions more complex than those which interest other horses. It had very little free time, what with one thing and another — always panting, partly to keep cool and partly to oxygenate its eight-times body. Its windpipe, after all, had only four times the normal area of cross section. And then there was eating. Somehow it had to eat, every day, eight times the amount that would satisfy a normal horse and had to push all that food down an esophagus only four times the caliber of the normal. The blood vessels, too, were reduced in relative size, and this made circulation more difficult and put extra strain on the heart.

A sad beast. (G. Bateson, 2002, p. 56)

Bigger, faster, better, cheaper . . . are actually none of those things. For example, industrial markets crave perfect produce, and agriculture has made bigger, more beautiful apples to meet that craving. These apples are ironically a symbol of health, not only the health of the apple tree but your health as well. Industrial agriculture has interrupted the apple tree's ability to respond to its environment to get such an apple into the market. Even in a drought, the trees are given water that does not go to other organisms in their landscape. Essentially, they rob their seedless, perfectly shaped juiciness from the soil and the other organisms. Their "health" is a trophy of linear, decontextualized exploitation.

They grow faster and bigger; they are Frankensteins. Whereas the little apple, the one with the dimples and the tough skin, is the one that shows it is responsive to its context. The drought the apple roots can sense informs the tree not to overdo it with extra fruit and growth. The big juicy apple is like that greedy jerk at the table who pours the whole bowl of raspberries on their plate and leaves none for the others. In turn, the preference and availability of the big juicy apple have obscured the apple-eater's ability to perceive the ecology of the apple.

Anything can be Frankenstein-ed. It is a danger of trying to hack a result out of an ecological process. Comfort and luxury often create a distance from the struggle of life. Like making perfect apples, the idea that life could be comfortable is a risk of not experiencing the suffering of life. Taking away that experience is not offering safety; it is, in fact, dangerous. We buy our way out of feeling and then treat the pain of separation with rapid-acting numbness. This is both extreme wealth and an impoverishment of communing.

Indeed, being in mutual vitality has to do with perception of others—perception with the others without othering. Perception without the conscious verbal mess. Perception of and sensitivity to the situations, without needing to be shown. And yet, the big juicy apple remains. When the wounds and numbnesses move, they are no longer self-confirming. More multiple transcontextual shifts will be outside the measure of the efficiency of the treatment, which is self-compounding trouble. If the pathology is identified as a disorder, and order is a path to destruction, there is a pathology in the pathology.

STUCK-NESS AND PATHOLOGY—THE ILLUSIONS

Over a decade ago, I had the opportunity to meet a group of people who run a clinic for paralysis and terminal pain in Italy. The "therapies" of this clinic include many seminal concepts from my father's work. I found their approach so inspiring that, to this day, I refer to this work in nearly every seminar I hold. I even brought the research team from the International Bateson Institute (IBI) to do a research project there. The clinic is called Centro Studi Riabilitazione Neurocognitiva di Villa Miari.

In 2012, the International Bateson Institute began a research project on "How Systems Get Unstuck." At the time, we thought the question was a good one as it addressed the stuck-ness we were perceiving around the world. The stuck-ness of the economy, the stuck-ness of the education system, the stuck-ness of the health system, the stuck-ness of the political and tech realms... mostly the stuck-ness of the thinking that would perpetuate such ongoing cultural stuck-ness.

It turned out not to be a good question at all, as you will notice if you think about it for even a moment. It is not a stretch to recognize that to have stuck-ness in a system, a great deal of compensatory movement is needed around it. To keep the market of fish at a constant rate, the industrial fishing companies have to work more and pay less. At the same time, they deplete the oceans of vast ecosystems and food chains that will expedite the degradation of the ocean floor and, with it, oxygen for the planet. To keep the economy at a constant Gross National Product, people around the globe are working in inhumane conditions, exploiting the natural world and the human community to get products to market for

Combining

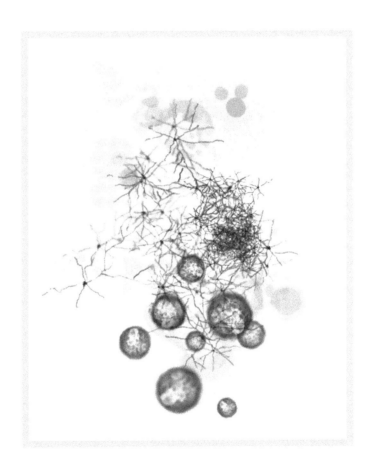

Fig. 40. Rachel Henstch. (2021). *Not Just That And Nothing More, version 1*. [digital art].

Fig. 41. Rachel Henstch. (2021). *Not Just That And Nothing More, version 2*. [digital art].

profit. A family member caught in an "addiction" will require family and friends to do all manner of social acrobatics to cover for them. In short, it takes a lot of change to keep things the same. Stuck-ness is only possible to identify within an arbitrary description of the edges of the system that is supposedly "stuck." If there is a stuck-ness, it is because one isn't looking to other territories of the system where the change is. It was not a good question. But we went ahead with it. Knowing we had a faulty question, we went to study paralysis as the closest we could come to a "stuck system." I will explain what we saw that opened the last decade of findings; it blew my mind.

The example I often share is this:
There are two people, one with a condition of paralysis or chronic pain, and the other is the so-called practitioner. (Note: These terms of reference already distort what I saw, essentially two human beings learning together. Identifying the two people puts them into an existing set of labels.) For lack of other language, I will call them the "practitioner" and the "person with paralysis." This language describing these two people is too significant to pass over without calling attention to its inadequacies and the habits the language ossifies.

The practitioner held up a small wooden block with a spiral carved into it and asked the person with paralysis to describe in words, without touching the wooden block, what it would feel like to move their finger around the spiral. The person with paralysis begins to verbalize a description. The description includes textures of the wood, speed of movement, temperature of the block, and so on. To make this description, the person is combining multiple forms of sense-making: visual, touch, temperature, speech, relationship with the context of the clinic, memory, and more.

Then, the practitioner gently and slowly moved the index finger of the "working" hand of the person with paralysis around the spiral. The practitioner's voice was soft and calm. After the finger was moved around the spiral, the person with paralysis was asked again to make a description. This time, the description was of the "difference between what they thought they would feel and what they felt"—which is an important distinction because, in attention to and articulation of that difference, there are three forms of perception coalescing. One is what was in the first verbal hypothesis, the second is in what was felt, and the third is in the difference.

Meanwhile, the practitioner seemed to be uninterested in the person's paralysis as such. The featured "frozen" hand and arm were not massaged or in any way pushed into opening. Usually, physical therapy would start with the parts of the body that were paralyzed. I wondered, "What "pathology" were these practitioners perceiving such that the response was making verbal descriptions of this little wooden block?"

So, I eventually asked, "What, in your opinion, is a pathology?" And the response was breathtaking:

A pathology is the organism's inability to make sense of its world.

At once, I began to comprehend the grace in this process. As these multiple forms of sense-making and experience were brought together, the whole person was learning in new ways. A transcontextual, mutual learning was taking place in which those pathways of cognition were re-forming through meeting the pathways of sight, touch, memory, and movement. The entire person—in relationship with their life history, the practitioner, and their language—was learning to make sense of their world.

Most importantly, the way they learned was unique, in their own time, with particular aptitudes and flourishing moments, followed by hard work and concentration. What was happening in the learning with the block enabled the body to release the blocked sense-making, allowing movement. But make no mistake, the person with paralysis did not "go back" to being unparalyzed; Instead, they were like a new people finding their way into a familiar but also new world. Who could know how each person's sight might unlock speech or where their memory of a particular minute motion in a hobby might give them access to feeling nuanced textures?

We were asking a new question: "How do systems learn?" This was the beginning of the concept of symmathesy, or (trans)contextual mutual learning.

For me, this was a portal into a study of how living systems are in constant mutual learning. How a crooked tree shows us the many ways it is learning to be in its world. It learns to shape itself, in-form, or inform around the other organisms, the wind, the shadows of other trees, and the soil. Similarly, a child who is raging or has gone silent shows us how they are learning to be in their world. A friend who takes a leave of absence to care for his aging mother is showing us how he is learning to be in his world. A grandmother who scolds her grandchild for not being independent is showing how she has learned to be in her world. This is non-trivial. The pathology was in the learning, and the healing was also in the learning.

What, then, is the difference between pathology and healing? Is there a difference?

> If double binds cause anguish and despair and destroy personal epistemological premises at some deep level, then it follows, conversely, that for the healing of these wounds and the growth of a new epistemology, some converse of the double bind will be appropriate. The double bind leads to the conclusion of despair, "There are no alternatives." (G. Bateson, 2000, p. 341)

Those particular connective processes must move in their own ways toward responsiveness with unity of senses, within unity of life, in many forms simultaneously. This is like symbiosis . . . this is like music . . . this is like the flavors of a soup combining—the ever-learning, ever-combining, ever-forming, and in-forming . . . nature never does one thing at a time. It is terrifying, and it is staggeringly beautiful.

> "Life learned early on to recognize itself." (Margulis & Sagan, 1995, p. 14)

This work is, undoubtedly, the most profound "system change" I have seen. In many respects, the Warm Data Lab is inspired by this work with pain and paralysis.

Combining

Did they heal? If healing is an ending, no: The learning continues. The body and the person must continue sensitizing to all sorts of physical, aesthetic, and emotional aspects of being alive. It does not end when the people leave the clinic. Each day, new richness of description continues to reveal itself. The learning moves and surfaces in unexpected ways.

When a person with paralysis comes into the clinic in a wheelchair and leaves on their own two feet, something like healing has taken place. This is a form of "proof," but there is a great loss in placing that language on what has occurred. It was so much more. The entire experience of being in a wheelchair is transcontextual; learning to be in the world changes, identity shifts, entirely different groups, friends, and unexpected lifestyles come into view. The world that the person is experiencing and what they are learning about being a part of that world is not just in them—it is in all of their relationships. Craig Slee importantly reminds us:

> For those for whom wheelchairs and other assistive devices have been part of life from the beginning, these devices are part of their extended selves and hence are sense and sense-making faculties, as much as, and sometimes even more so than the "traditional" attributions and organs. So, it is perhaps important to note that for many, even framings of ability or inability raise yet more questions. (Personal Conversation, 2022)

So, the walking is not what is important—for many, the healing will take place around being in a wheelchair in a full life. It is not just about feet and balance and momentum; it is about a perception of the world as it shifts visually, emotionally, intellectually, feeling texture, and so on. One metaphor might be that when they arrive at the clinic, parts of the world around them are invisible, numb, flat, empty. Then, they gradually begin to sense, with many senses, a world of texture and dimension filled with memories and possibilities.

The world was there; they were out of range for a while—sensitizing . . . and re-sensing the world so that it is possible to experience it from many perceptions becomes integrity, generosity, and learning. Healing is mutual in that sensitivity—and ongoing. For me, this is a very real way to describe the undercurrents of much of the pain and stuck-ness of our world. It is an invitation to approach this with a similar elegance of indirect connectivity of senses—allowing for the differences of experience to rub against each other—allowing memory to find new pathways of meaning-making. The responsiveness to the world around us is only as perceptive as our organisms can muster. What makes this so beautiful also makes it difficult to make a place for it in our existing understanding of the movement from incoherence to grace. The paralysis is not removed—it learns.

What can we do to allow this sort of sensitivity to find us? And when it does, if we call it "healing," will it be smothered in a strategic manipulation that can be made into a methodology, a treatment program, or a therapy? If it is not "healing"—what is it? Can we keep it outside of this form of thinking? Is it possible to name it without it becoming metabolized by the habits of productivity?

As one participant, Liyana Musfirah, so beautifully describes it:

> Warm Data Labs provide a blank canvas for the new soul who is uncertain and unsure of what should be on the canvas. Upon entering the realm of Warm Data Lab, the soul couldn't help but wonder on the kind of colors, the textures, and the themes that should be on the canvas. It is a blank canvas that allows any soul to come in and paint the sincerest and most raw of pictures. But one needs to realize that these pictures, these new colors, and textures, these new themes, they are not merely art; they carry experiences, they carry stories, they carry history. And what's beautiful is that these pictures carry life. These canvases are now alive. They are alive because not only one soul paints on the canvas, but rather many souls paint the canvas as the same time, so the canvas keeps breathing a new form of life over and over again. This realm offers any soul to connect and interact through the multiple layers of colors, textures, themes, and histories, so it is not simply a realm, it is not simply a canvas, it is an ecosystem that allows new life to form." (Personal Conversation with Social Media Influencer, 2022)

As my father said, "I hold to the presupposition that our loss of the sense of the aesthetic unity was, quite simply, an epistemological mistake." (1979, p. 17)

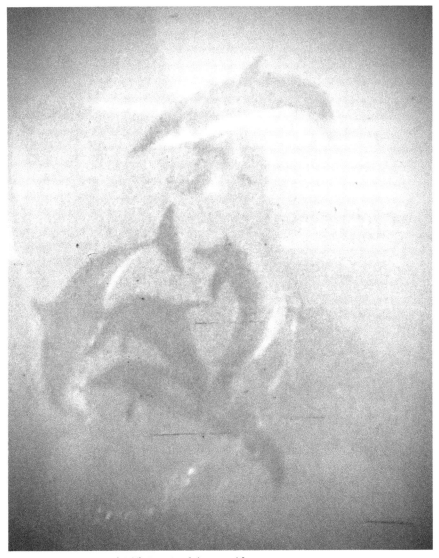

Fig. 42. Gregory Bateson. (1966). *Porpoises*. [photograph].

Unbreakable

"If I told you that a flower bloomed in a dark room would you trust it?" - Kendrick Lamar

The revolution, the evolution, is not going to be found in conference centers or seminars in five-star hotels. It won't be definable in righteousness or sanctimony.

The resonances will be and are where the pain has been — where there was no choice but to become unbreakable.

Where the scars are generations deep and sculpted into raw, sassy, funny, sexy, harsh, confusing.

Laser sharp, intelligent literacy of metaphors; the messages won't be sterilized in direct instructional language. The (r)evolution is not a Lego set to assemble; it is alive.

The alchemy of change changes as it changes.

The logic of the current systems is not so useful for the task. It leads back to the academy, city hall, and the bank. Those are not the places where unbreakable is made.

Look elsewhere.

WHO-NEW?

The unfamiliar is always going to appear
incoherent.
It won't make sense.
To what we thought we knew,
At first.
Then.
The importance of surrealism—
Reminds us also,
That stranger times
... would offer the new.
But who would know?

WHAT AM I NOT ABLE TO RECEIVE?

Differences in history and culture appear to create the need to tailor communication and relationship to authority differently. This communication needs to contain familiar recognizable information receivable within the culture. The same signal is received differently in different contexts at different times. For example, if a particular state tells a citizenry of rugged individualists what to do, they are likely to rebel. Other populations expect clear instructions and honesty. But those things look very different in different cultures. And with lives on the line, the need to mind this cultural signaling is serious.

Living in a culture that is not my own, I have to wonder, "What am I not able to hear?" This has been an informative discernment. I am not able to pick up on the deep contextual and historical root systems of meaning or the attitudes and responses of cultures I have not lived within. What is totalitarianism to some is safety to others. What is freedom to some is reckless to others.

However, I have also noticed that in each place, there appears to be something like cultural scar tissue — where old, collective shame, wounds, and numbness are currently generating denial and defensiveness. These dark spots seem to hinder each culture in their response. The double binds of the past, the justifications for things that should not have been justified — these old wounds are guarded most fiercely ... and present most readily as culturally specific notions of why the havoc and loss, from the polycrisis "won't happen to us ... "

And then it does. And the scaffolding is a little more visible with each crack that reveals its invisible edges.

COMBINING

SURREAL

This is not a hidden language.
There are holes in the world arranged in
reaches.
Where unstrung guitars are strummed by
untethered players performing unbound songs.
Nothing is pre-made or full-service —
except the stretching.

Maybe I am listening to a room full of apples,
While curtains drape like forests in midair,
Clouds step in —
Breaking the frame of what once was Magritte's
black hat.

With articulate distortion,
Invisible paradoxes are difficult
conversations,
The art is in memory's imperfections.

Occasionally a new portal arrived.
Important new ideas.

But here we are
Still whispering underwater.
Growing gills.

Fig. 43. Vivien Leung. (2023). *Belly Spinning*. [digital art].

Decontextualized

Fig. 44. Alexei Jawlensky. (1917). *Mystical Head: Galka*. [Image of oil and pencil on tan textured cardboard]. Image from the Norton Simon Museum, The Blue Four Galka Scheyer Collection

Fig. 45. Ursus Wehrli. (2004). *Kunst aufräumen*. [Image of painting]. Copyright © Kein und Aber Verlag Königstein i. Ts.

The red and blue brush strokes (to the left) combine into the viewer's collection of life experiences to form cheek bones, melancholy, and personality. When they are categorized and stacked (as above) the possibility of data combining is completely different, and it is impossible to see the woman again.

Imagine each stack of colored brushstrokes below as a context of identity defining the person by statistical info. For example, economy, legal, health, education, technology... the warm data is gone. The meta message of the stacks is one of control. In contrast, the meta message of the painting of the woman is of life.

Where is the person in the stacked colors?

A List of Relevant Questions

The world is in trouble. There is serious work to be done. Quickly — what follows is my shortlist of serious questions for this era. Structural change is required. To get there requires epistemological change. This list assumes that profit, revenge, ownership, and any notion of superiority including nationalism are not viable excuses for the destruction of billions of living systems. The committees that are formed, the projects that are funded, the actions taken, the policies made — all are in response to the questions posed. So what are those questions? Here are mine.

All of the questions below are of personal, institutional, and global concern. "We" in this document refers to anyone who will ask such questions with me.

Education: How can we best cultivate curiosity, information, and learning between generations to prepare ourselves to perceive and respond to the complexity of our world with less destruction than in centuries past?

Health: How can we support health in human beings by making it possible for each person to eat healthy food, sleep well, know that their families are supported, be respected in their community, have relevant contributions (education and employment), breathe clean air, and drink clean water?

Ecology: How can we interface with the complexity of our natural world so as to cause less harm to the interdependence of all living things?

Economy: How can we shift the economic system so that it is not based upon exploitation of nature and humanity, without crashing the globe into chaos? (note: no one gets rich on this version of economy)

Politics: How do we get the policy makers of our world to mandate cross-sector information for their decision making processes so that they have the possibility of taking into account complexity?

Media: How do we get a moratorium on binaries? How do we support public understanding, not trained in perceiving complexity, to become accustomed to it and demand that communications institutions deliver cross-contextual information?

Culture: What is the approach to open the global discussion about the pending fate of humanity? What matters? What are we willing to change? How can we survive together?

Transcontextual

Transcontextual

"Let me coin the word "transcontextual" as a general term for this genus of syndromes.

It seems that both those whose life is enriched by transcontextual gifts and those who are impoverished by transcontextual confusions are alike in one respect: for them there is always or often a "double take."

A falling leaf, the greeting of a friend, or a "primrose by the river's brim" is not "just that and nothing more." Exogenous experience may be framed in the contexts of dream, and internal thought may be projected into the contexts of the external world. And so on. For all this, we seek a partial explanation in learning and experience."

(G. Bateson, 2002, p. 277)

Fig. 46. (pp. 308-315). Rachel Hentsch (design & concept), with Nora Bateson (Concept & Text) and Leslie Thulin (Concept). (2023). *Transcontextual*. [digital art].

Law - Family - Health

CONSENT
"Go for it" - Motto of the modern world.
"Just do it" - Nike slogan

IMMIGRATION
Legal systems are (also) struggling with increasing pressures to harmonize with international norms that differ from local traditions.

ADDICTION
Clearly the sale and use of prescription drugs outside the dosage given by a medical professional is illegal. So are the peripheral petty crimes and more serious criminality that come with drug culture. Additionally, the malpractice and other lawsuits against and between practitioners, treatment centers, and other legal snags, represent a vast body of legal work.

Criminality is often the first cultural stigma that gives imagery to the topic of drug addiction. This criminality suppresses communication of the systemic societal issue that is emerging as it effectively gags families, doctors, and other people whose lives are changed by addiction.

IMMIGRATION
Colonialism is not a thing of the past.

ADDICTION
When one person in a family, community, or society is suffering from addiction the entire network of relationships in that individual's life is affected.

CONSENT
Each family has a culture of its own, scars of its own.

SCHOOLS & INVESTMENT
What is it like to be young in a moment when the shared vision of the future is filled with upheaval? What dreams, what stories, what forecasting can children today hold onto? Will they follow the path of the generations that came before them?

IMMIGRATION
Between generations, at the dinner table, in the garden or the market, this is where ideological threads are woven into the ecology of institutional systems. Families are where attitudes toward other people and nature are expressed and passed on.

Education – Politics – Ecology

ADDICTION
Educational institutions have a responsibility to become informed about how drugs, that are commonly known as "study drugs" and prescribed as ADHD medication, are incorporated into school culture. At the elementary school level, the diagnosis and treatment of attention-related issues need to be put into the larger context of the long-term health of the child.

INVESTMENT & SCHOOLS
Education that takes place in nature stimulates entirely different sets of cognitive processes. Wilderness as a classroom is the original laboratory, playground, and library of organisms in interaction.

CONSENT
Gender roles and expected behaviors are woven into the classroom experience.

CONSENT
The law cannot effectively meet the challenges of consent. Rape and abuse accusations require evidence, which is often impossible to provide.

SCHOOLS & INVESTMENT
What is learned about social engagement and responsibility in school? With the rise of populism in the theater of international politics in recent years, it has become apparent that there is a need for a citizen base that demands more than binary positions.

IMMIGRATION
Democracy is under-equipped in its current form to accommodate the complexity of the issues emerging globally.

CONSENT
Attention to destructive patterns of exploitation between people may help make visible the exploitation by humanity of the environment.

INVESTMENT & SCHOOLS
There is an urgent need for an education system that offers the conditions through which perception of the complexity of the environment, as well as of each other, is possible. It is a mistake to think that young children cannot see, or work with, complexity.

ADDICTION
The political implications of legalizing and making available medications has a complicated history in terms of culture, morals, and money trails. The ongoing developments in research also influence this process. The dual economies of the pharmaceutical and black market sales of drugs both create and sustain political involvement.

ADDICTION
While it might seem that issues of addiction are limited to the parameters within this culture there are patterns within the drug crisis that relate to the larger societal issues. Addiction, like ecology, is formed through a combination of patterns of living and cognition that together become inextricably interwoven.

IMMIGRATION
The ways in which ecological changes will affect culture and economy are yet to be seen.

Economy - Culture - Technology

CONSENT
What are the economic repercussions of denying consent?

IMMIGRATION
Information about complex topics like climate, global economy, and the refugee crisis is rarely communicated in its full complexity over mainstream media and journalism. Mostly stories that have more than two sides are regarded as complicated and unsellable.

SCHOOLS & INVESTMENT
The idea that there are right and wrong answers is a cultural binary that undermines inquiry into complexity and diverts us towards problem-solving attitudes throughout every sector of society. Misunderstood complexity often results in an ill-informed need for solutions, strategies, concrete deliverables and other products of linear thinking. This approach is perpetually confounding those needing to grasp, and deal with, complex circumstances.

CONSENT
The media provides the conduit through which many unspoken assumptions are conveyed. Consent is woven into ideas of dating and into love stories.

IMMIGRATION
The most familiar line of discussion around the refugee crisis is that the arrival of people in need will disrupt and perhaps destroy existing economies.

INVESTMENT
If extraction and exploitation become impossible to continue, the entire basis of profits and markets will be disrupted. What will be the skills that will be relevant in such a society? While this may seem far-fetched, it is a question worth pondering.

CONSENT
The mechanistic efficiency of technology seems only to increase. The "next gen" stuff is always faster, smoother, and trickier than the previous model.

ADDICTION
What does a successful life look like? The purveyors of the imagery of success and happiness are largely cultural. The often unspoken assumptions of what is normal, what is expected, and what is respected are potent messages about how to live. Skinny, beautiful, rich, successful, fit, perfect people are plastered all over the media.

INVESTMENT & SCHOOLS
What are the similarities between history and algebra? This sort of connection may seem unusual to consider at first, but not impossible. Both require that students think in terms of variables.

IMMIGRATION
Human beings are creatures of place, and our identity, belonging, and well-being are woven into landscapes that house our history.

SCHOOLS
The metaphors that run through the culture inform the collective in ways that are so deeply rooted that they are hardly visible. Presumptions regarding authority, dominance, status, and even the notion of what health and happiness should look like, are culturally scripted into daily life.

ADDICTION
The cost of responding to widespread addiction issues as they crop up in their varying forms is astronomical. From treatment centers to unemployment, from overdose care to automobile accidents, psychological treatment, legal fees, prison costs, foster care, education and so on. Addiction is expensive to address, and the expense crosses many sectors.

CONSENT
What does it mean to be desirable? Where do these ideas come from?

Family is Where We Live

Family is the blood tie and the non-blood tie. Bound, unbound, and always returning to the tension between coming and going, caring and abandoning, living and dying. Family is a time-boat that spins in a whirlpool of present, future, and past. Ancestors sing and spit; unborn generations write their inspirational notes in treasures and traumas they will inherit. The days pass, and family is the soup of it. Family is sour, savory, and nourishing, while it is also impoverishing.

Family is all of it—the close ones we love to care for and the dangerous ones. Family is a raging sea, a rugged mountain, a warm fire, and an orchard of old fruit trees that have fed both grandmother and granddaughter. Family is acidic and cruel; the sibling that would kill you, the uncle that would rape you. Family is anyone that is there for you at four in the morning. Whether or not someone shares your DNA, or your bed, or your household, or your history or your money—family is an ever-shifting ecology.

Where is the edge of your family? Is it parents and children? Is it the microbiome and the ecology? Is it the culture? Where is family in the history and hidden stories of love and destruction?

But family is not what it used to be. No nostalgia, just paradox. Family was always fraught with contradiction. People have always toggled between needing and not needing their families. We have all relished and resented giving to and receiving from our families. The familiar, which even has the word "family" in it, is not a static state or a predictable pattern. The familiar in a family is the reality that each family member is changing each day. Each morning, we wake up both knowing and not knowing our family.

Family is where we live. Not in countries or careers, or sectors or economic classes . . . in families. Even if I live alone, my aloneness is written into the endless stretchy tendrils of my family.

This is relationship in complexity. This is day-to-day living.

The next decades of human life on this planet will reveal how each of us has nourished the relationships we live within—today, tomorrow, and next week. Did we perceive the complexity in ourselves, each other, and the world? Did we tend to the relationships that build more relationships? Attention to words spoken to the child that set the tone for the grandchildren's stories—the gift of a weekend conversation with the elders that holds a meta message of care for the past and future—willingness to be wrong, willingness to be angry, willingness to be on the swirling time-boat together…

For me, the big question now is about how relationships are approached. What is essential? What is the grounding of the relationships? Whose script are we on? What is it possible to say? What is it NOT possible to say? Who is it possible for me, or you, or anyone to be in this ecology of family? Are we playing the games of who is right and who is wrong? Is this a big stage on which to debate the past? Shall we dip backward into the formality of communication rooted in the legal system? Will we hold each other to the metallic edge of a mechanistic culture that justifies without context? Will it be asked, "Who said what, is it provable, what is the truth?"

Or will it be obvious that in each moment, the contexts and histories of each person bring their own reasoning? When contexts are perceived, direct correctives are vaporized, the interdependencies are seen, and the response is of another order. Exhale. There is no judge, only the deep waters of context and the unexpected insights that life provides. There is no script for what comes next, only practice in the stance of perceiving and responding to the complexity.

Relentlessly method-less, map-less, label-less—the possibility for systems change lies in the connective tissue of how each of us is family-ing. The entire human species depends upon attention to finding art forms of mutual learning. The generations will learn together how to live differently or not learn at all, which means that authority has to be a permeable membrane that can learn and be learned from simultaneously. What better way to show the younger generations how to learn than to let them see us learn from them? What better way to enter the coming decades' challenges than to practice improvising life together?

Family is a double bind that will both trap and set you free.
That is how it is. And, probably, how it always has been.
Family is a time-boat.
… on which we will give birth in the high winds of change,
to another form of loving each other.

"If a thing loves, it is infinite" ~ William Blake (1788)

---

* Previously published in *The Challenges of Today's Families: A Systemic Approach to Family Relationships*

Fig. 47. Leslie Thulin. (2023). *Weaving*. [digital art].

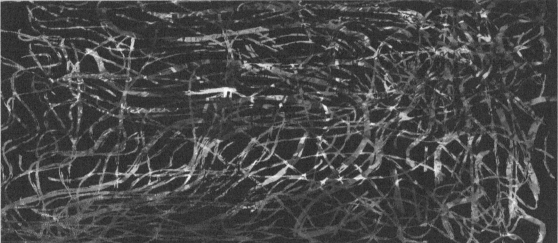

Fig. 48. Leslie Thulin. (2023). *Scraping*. [digital art].

Fig. 49. Leslie Thulin. (2023). *Cracking Frequency*. [digital art].

Meeting Double Binds in the Polycrisis
Starting at Home

The double bind theory describes a pattern of relationships that appear to create only painful traps. The bind is formed through multiple contexts together, producing a situation in which success or solution seems impossible. Additionally, familiar forms of communication cannot be used to respond to or even describe the double bind. The double bind is a contradictory relational situation in which it is perceived that only failure is possible. My father and his colleagues at the Bateson Group introduced the double bind theory in their paper related to schizophrenia in the 1950s. However, my father had been working on the theory long before that publication, and he had always intended it to be a theory of evolution, learning, and metaphorical communication rather than the causality of any singular "human pathology." I often say in my courses that if there were one theory I wish I could share with the world, it would be this one. While recognizing a double bind does not remove the bind, it opens possibilities that might otherwise be imperceptible. It helps. It is worth it if the pain of experiencing double binds can be relieved a little bit through familiarity with the concept.

In this era, simultaneously encountering multiple double binds is an everyday experience of unfurling interlocked catastrophes through multiple contexts of daily life, including economic crises, ecological crises, political crises, technological crises, immigration systems, education systems, food production, health systems, and wars. Polycrisis is the term often used to describe the phenomenon of multiple concurrent global and personal emergencies. Understanding double bind theory is a necessary first step toward creatively meeting a polycrisis.

Double binds are everywhere—at home, in the workplace, in education systems, in spirituality, in health systems, in the economy, and even in natural forms of evolution. There are times when the context an organism lives within necessitates particular forms of behavior for survival, and then times when those contexts change. When this happens, old behavior may become fatal, but the organism only knows how to survive in the way it has in the past. Previous survival behaviors will prove deadly, and new behaviors will appear also to be a deviation from survival. Addiction is also an example of the double bind: the person with the habit may know that the substance and/or behavior is destructive, but the idea of living without it is as foreign and unthinkable as being asked to suddenly grow gills or a new limb; it can be terrifying.

The double bind can be brutal, tormenting, and devastating to the person or organism involved. But the double bind is also a creative imperative. It is the moment when an organism or a person must find a creative jump to a new understanding of the relationship to their contextual surroundings, which may sound easy, but it is not. The pattern is called a double bind, but the binds are more than double; they reach deeply into the ecology of the situation, the relationships, the communication, and the histories. These tangles shape more binds over time; they get deeper and more convoluted as the loops repeat, creating new binds on top of old binds. The overlapping and knotting make it feel impossible to escape from the pulling of the binds as they tighten their grip. The result of not making the jump is obsolescence. But often, making the jump to a new perception requires letting go of an old understanding. The experience of the double bind is of a pending death either to identity, a group of familiar perceptions, relational dependencies, or physical survival. Transformation is as creative to new ways of being alive as it is destructive to the prior ways of being alive. Finding new forms of living is not about making an adjustment, or tweaking existing structures; the jump is transcontextual.

The daily news reports and formal analysis of polycrisis events—like the uncertainty of economic stability, increasing numbers of unhoused people in urban areas, loss of insect population, and melting glaciers—are communicated in a tone of objective abstraction and perhaps a call to political action. This public media form of description of the polycrisis can feel distant and out of reach of most people's field of influence. However, there is another tone of overwhelming frustration, despair, and loneliness in our homes. People have their own versions of complex system breakdown in one form or another, living intimately within the polycrisis, some feeling the catastrophe more than others. Most people do not recognize their pain as an expression of a polycrisis. Households currently hold multiple stories of depression, anxiety, loss of income, addiction, and myriad physical and mental health problems. Instead of seeing these issues as consequences of inter-systemic breakdown, people are taught to either assign blame elsewhere or to blame their own "bad choices" and lack of willpower—thereby perpetuating the illusion that the individuals are "broken" while the system is fine. Placing the blame on individuals further isolates people from their potential perception of the systemic polycrisis that they are co-existing within, deepening their frustration when social services cannot respond adequately. Seen through the lens of the polycrisis, we can see that these stories are produced through multiple contexts and require a multi-contextual response from inside the community. Many of the traumas people are attempting to heal are, in fact, located in systemic transcontextual issues of the multi-generational build-up to the polycrisis of today, even though the pain is felt to be personal and labeled as located in the individual.

Now, the increasing number of people in need of care and treatment is surpassing the capacities of the service organizations that are supposed to be able to respond to these issues. Additionally, the issues have been overlapping for decades and over many generations. It is rare that those seeking help display just one symptom. A single household may be simultaneously dealing with abuse, poverty, addiction, cancer, and mental health issues. These expressions of the larger global polycrisis are unlikely to be revealed as lone symptoms. Yet, the institutional, procedural protocol will parse the issues to different departments; there is no authorization to treat abuse, addiction, poverty, and illness in the same office. The organizations authorized to offer care are structured within the same systems that generate the conditions contributing to people's suffering. For example, it is not uncommon for teenage patients to be assigned a different doctor for multiple diagnoses—an eating "disorder," anxiety, kleptomania, and ADHD. Many times, each doctor prescribes their treatments separately, and in some countries, it is forbidden by law to communicate treatments with the other caregivers. This siloing of personal, familial, and communal crises has placed the social care systems in an impossible situation. The rigid structure of the systems that deliver care is also caught in double binds, which causes more impossibilities to emerge.

A visual description of this categorized structure can be seen in the United Nations Sustainable Development Goals logo. The logo depicts seventeen brightly colored tiles, each labeled with a singular ecological, economic, or cultural/political crisis. There is one for poverty, hunger, gender equality, life on land, life below water, quality education, reduced inequalities, and more. Though named separately, it is widely acknowledged that these goals are entangled. However, the perception mishap in the logo is missing more than the interconnection. I invite you to envision another image of a mother nursing her baby and place this image side-by-side with the grid of colored tiles—both are metaphors for communicating the continuation of life. All communication occurs in multiple contexts, some explicit, some implicit, some associative, and some tacit.

In the metaphor of the mother nursing her baby, every single one of the Sustainable Development Goals is there, not broken into a grid, but integrated into one of humankind's: most life-giving images: intergenerational care and nourishment. The symbolic message of mother and child is biological, emotional, intellectual, ecological, economic, and cultural care. To continue our species, human beings, like all mammals, must ensure that the next generations are fed. The single image of a mother feeding her child includes the mandate for clean air, clean oceans, gender equality, education, and so on. While not everyone is or will be a parent, the future is held by the next generations. For all parents to feed their babies, the people who grow the food and make the clothes must live in a world where they can feed their babies, and their babies must be able to feed their babies for generations to come. Feeding the babies also addresses the crisis in more than a 1st-order direct response—it meets the future needs of humanity in the nth-order. A child who is cared for and loved can keep giving loving care to family, land, and community in many unforeseen ways. The polycrisis, a consequence of the consequences of so many double binds is best met (not matched), with support for and in the home, where all the crises and possibilities come together.

When a child is struggling with school, when a parent is struggling with an alcohol habit, when a family member is injured or ill, when someone you care for is unable to "successfully" participate in the socio-economic structures demanded of them—it is the family and loved ones who absorb the stress. The home and those loved ones have been giving time, care, housing, money, and love that created resilience in communities, invisibly making up for the consequences of the systemic stresses of day-to-day life. With the increasing tensions of the polycrisis, there is a decrease in resilience in intimate relationships. Too many people are beyond their capacity to respond with care to their own pain and chaos, let alone to be able to help their friends and family. If one member of a household is unable to participate in day-to-day life due to injury, illness, mental health, substance habit, or employment loss, people close to them will be there to offer emotional, physical, and even financial support. Even though struggles and stresses are met in the first response by members of the same household, the need to nourish this nexus of resilience is out of reach of individuated labels and treatments. Without the solidarity of communal mutual care, there is a significant loss of connective tissue to entire communities, leaving households and individuals susceptible to many traumas. A new, warmer logic of response is needed to meet the interdependency of crises. The image of the mother nursing her child illustrates how natural systems of nourishment meet multiple needs in a single gesture of love and communication.

For this reason, I am attending to the double binds of social care and the dangerous, systemic breakpoint pressures they generate. In these times of collective trouble, caregiving is crucial. The resilience of the family, neighborhood, and community entirely depends upon the strength and health of people living together. Because this kind of care has no definable boundary or measurement, it is largely unfundable, unresearchable, and un-tendable.

A tragic example of the double bind is seen in the way "work-life" produces stresses that undermine family, community, physical health, mental health, caregiving, and all living systems. Most people need some form of income to buy food, shelter, and basic needs. Additionally, material wealth is wrapped up in identity—especially in the measurement of success by which people assess their sense of worth and each other. Often, the first question people ask when you meet is, "What do you do?" If you cannot perform in a "professional capacity," there is an underlying message that you are a misfit or a cost to the system—an expense—you are not doing "your share." This incredibly violent way of perceiving people hides the gifts they have to give because those gifts do not fit into the "workforce" as it is now (or the idea of a workforce in the first place.) Artists, children, caregivers, older people, gardeners, people who have been injured, those who are ill, people who want to change the systems, activists, people with mental health struggles, people who have been born into communities bound by inter-systemic racism and excluded from opportunity, and so on—are all un-placed in the existing categories of the grid. Also, those who are in the grid are incurring enormous damage from being in the grid—the destruction then spreads to other relationships—it may be the single parents who must leave their children during work hours, the family members of people who need care but who cannot give the care, and the adult children who wish to assist their aging parents but cannot disrupt their work. When one person in a family or household is in need, everyone else must bend to the situation in one way or another. The stress stretches into the lives of partners, children, and friends of the family members who need to give care to their relatives but cannot skip work, and so lean on their significant others to help with the burden of timing and finance to be able to deal with the pain of not being there.

The machine that is the workforce is also bound by national issues that loop. Social services supported by taxation make social welfare a tangled idea, suggesting that some people are "paying" for those who do not fit in the workforce. For those on the receiving end of the social services, there is an open door into rippling shame toward losing personal esteem—piling still more obstacles on people already marginalized. The signals of both material wealth and poverty allude to how the existing systems have shaped these contrasting identities—whether someone is driving a flashy car or holding up a sign asking for help. While current institutions and workforce configurations appear necessary for survival in our current state, it is also deadly. People are falling through the cracks and becoming increasingly unable to bear the stresses of the isolation and mono-cultured days of repeatedly doing the same thing. The daily grind is a double bind. The answer to this double bind is not recovery programs to fluff up motivation and strength to re-enter the same systems draining vitality. Every institution is rife with double binds, all authorized by a pervasive industrial logic. Each office is limited to its forms, categories, and procedures. Everyone is caught in double binds—social workers and those in need. Even with a clamor for change, institutions cannot undo themselves.

COMMUNICATION AND METACOMMUNICATION AND THE SOCIAL SERVICE QUANDARY

Central to the double bind theory is recognizing the importance of metacommunication. Direct communication can be pointed to as what might end up on a transcript (if there were one) of any interaction. However, what is said is not what is said—what is said is what could be said in the ways it was possible to say it. The limits around what can be said are not direct and are within the metacommunication. In a social service context, citizens may be informed of their right to receive care, but metacommunication dictates the logic and form of that care. It is clear to most people entering a care facility that they may receive relief from the pain they are in, but the facility cannot undo the conditions that are creating the pain in the first place. This limitation is communicated, yet it is not written anywhere. Likewise, the metacommunication in a governmental building will transmit that no care for psychological or physical pain will be offered within those walls.

The metacommunication within systems of facilitated care limits both those who can be defined as in need of care and those who offer care. Both are in communication through the expectation, tone, and logic of the legal system, the school system, and the health system. The logic of these systems is coherent to itself, and evidences its own productivity within a culture of metrics shared across social systems. Results of individual improvement or failure are provided as indicators, again forcing the description onto the individual. Each system operates within a doctrine of efficiency, objectivity, and labeling; all achieved through reductionism. The message is laced into the walls, the desks, the forms, and the questionnaires of institutional caregiving. The person who seeks care is likely to be aware of these limits if they have any experience with other institutional offices and will communicate within the correct boundaries of information disclosure. Those who do not understand the system will likely try to communicate contextual information the offices cannot take in. There is no form for the contextual and transcontextual information of several hundred years of double binds. In the metacommunication, there is an expectation that the person in need will describe their issue as singular and seek singular, predetermined treatment for the symptom. Understandably, the sector has had to operate within policies that would secure the safety of patients and caregivers, which required a flattening and distancing of relational possibility to keep it from getting messily over-personalized. However, this is a zone worth exploring as too much information is currently being excluded from the relationship (of caregiving and care-receiving) by the system logic of reductionism. Data generating is compounding many of the double binds by eliminating contextual information. Should single mothers living in their cars be counted as cases of diagnosed depression? While it is natural to feel depressed in their situation, the diagnosis re-directs the perception of the situation.

Services, be they social services, spiritual retreats, or self-help programs, that are tasked with returning supposed "broken" and burnt-out people back to "normal" are essentially re-tooling us to get back into the workforce (and social roles) that ate us up to begin with. They are assigned to fix the people but not to restructure the system, undermining the family, the natural world, and hope for the future. The social help systems perpetuate the systems that produce the need for social health systems. The exploitation and objectification of people's lives inherent in the economic stability of the national and international markets are inherently destructive to life itself. Similar decontexualizing is reflected in the way services are assigned by category of care needed, such as hunger, abuse, loss of housing, childcare, mental health, physical illness, bankruptcy, and so on. While these issues are all formed in multiple causal processes, the interdependency of these processes is not within any department. Single parents living in their car with their kids are offered antidepressants, even though the medication will not help alter the dangers of the wealth gap that produced their lack of housing. I believe this is as upsetting for the social workers as it is for the parent and, subsequently, their children, who may experience a lack of responsiveness in their parent due to the meds and, in turn, may begin to act out at school . . . only to be given another diagnosis.

In the case of mental health and often physical health, a diagnosis is required before services can be given. The history of medical care, psychological and physical, is rooted in diagnosis. This practice has offered developments in care, and it has also obscured systemic healthcare. Today, diagnosis is the communication that the existing system of care can respond to by compartmentalizing and decontextualizing. Likewise, the medications and therapies will often be singular in their treatment. The symptoms will be treated, and the multicausal processes that have produced the problem will be largely ignored. For some people, treatment of the symptoms will offer relief. Having a name for the problem is the beginning of not feeling so alone and unable to fit into day-to-day life. Some feel seen by the diagnosis. For others, the diagnosis makes their situation unseen, as it presents looping issues of being identified in one category when the struggles and suffering being experienced are spread across many categories. In either case, likely, the person cannot receive any care or dispensation for work loss if they do not present a diagnosis that justifies the need. Once assigned a diagnosis, the records may later be considered against you in a job application process. The diagnosis forms its own double binds—without the diagnosis, it is difficult to get treatment, but with it, the treatment becomes dislodged from its complexity.

A good friend summed up the workforce/caregiving double bind in a single sentence: "I took a leave of absence to take care of my aging mother." This statement vividly reveals the double bind that afflicts all caregiving and life-affirming and nourishing processes outside the workforce. In the case of my friend, he had the financial flexibility to take time off, but most of us do not. Most of us would lose our homes, jobs, connections to colleagues, health insurance, social position, and maybe even life partners if we got off the wheel. What does it say about a so-called "society" if caregiving is not valued? Raising children, tending to the elderly and the sick, and even tending to the ecosystems are seen as expenses, not vital. What does it say about a system if all the necessary activities that produce intergenerational health and well-being are outside the economic metabolism?

The double bind of the workforce is this: If I don't work, I can't live, and if I do work, I can't tend to life, and my physical and emotional life will be damaged or destroyed. There is nowhere I can submit a complaint form to address this. If I break, burn out, or collapse under stress, at best, I will be given either medications or other therapies to put me back into the system that is breaking me as soon as possible. There is no getting out of this hoop. Exiting the system is not part of the system.

How Do You Get Out of a Double Bind?

To meet a double bind is not to try to solve it from any one direction. Instead, to meet a double bind is approaching it from another context, usually one not perceivable in the immediate struggle. As I mentioned above, this is not easy. The perception of the bind is loud, and it screams in polarities, dragging and pulling us through binaries. The addiction is perceived as a quit/don't quit scenario. The rent is seen as a pay/don't pay scenario. Caregiving is seen as a care/don't care scenario. Everything appears to underscore the stuck-ness of the double bind, creating a desperate perception trap, which is the nature of double binds and why it is so hard to get free of them. The possibility of another perception is not perceivable through the clamor of existential trial.

To meet the double bind is not to match it. The necessary response must come from another context besides those inflamed in the bind, with an entirely different aesthetic, tone, feel, and logic. The response will not be flavored with the metallic measurements of solutionism. It will be weird-looking and seem surreal; it will be off-topic. It will not be accountable to the departments hooked on decontextualized information. The metacommunication must be inclusive of this shift in possibility. What is possible to communicate, put on a form, or be asked in a questionnaire must be expanded to welcome the unfamiliar and uncontrollable realms of information that, up to now, have been blocked. The context of social caregiving must include the combining contexts of living in an era of polycrisis, which will alter the feel of the profession—from the walls of the offices to the voicemail options to the possible ways to help each other. The tone of efficiency and reductionism itself is a metacommunication holding the limits of how people can even begin to think about their double binds, let alone respond to them. Unless there is a change in the aesthetic of communication in these services, no structural or policy-based alterations will open the double binds. This shift in metacommunication is also a shift from abstracted labels and distant professional "objectivity" to curiosity, humor, play, ritual, poetry, and story within the mutual learning between caregiver, patient, and community.

The meeting of the double bind is radical, not because it will be seen as activism, but because it will not even be recognizable as a response to the crisis at hand. Addictions and habits may be met through recognition of responsibility in another context; the stress of work/caregiving may be met through community. The details in the contexts to the side of the double bind will often reveal an unexpected possibility. It's not possible to bring this information in when the crisp lines on the form do not allow for new, transcontextual movement; the information needed to meet the double bind is not static or compartmentalized. An entirely different tone of approach is required. What might that look like in terms of creating methodologies for social services or institutions that care for health, education, mental health, older people, youth at risk, addiction, and so on? Is it possible to generate a manual for care that breaks through the double binds of our world? This vital question highlights the issue at its core.

Combining

There is a common-sense assumption that a method or a manual will be the first thing to attempt to address the defined "problem." But, this is precisely the thinking that is creating the issue to begin with, which brings us to another double bind—the double bind of how to fix social services. If they are not changed, they will continue to generate perpetuation and justification of the existing systems, which are harming people and the earth. Still, if they are changed, they are only changeable in ways that make sense within a system that justifies and perpetuates existing systems of harm.

As these systems increasingly lack sufficient funding, it is becoming clear that for most communities in need in decades to come, no one is coming to save them. There is not enough support for this failing system to meet this era of polycrisis. The needs are too many, and the associations licensed to give care are too few. A new logic is needed; it will not be confirmed by boards of directors or their by-laws. It will not be coherent to the voting public. It will be wild.

Wild as in alive, in the moment, and able to meet the complexity of each situation. Each family will need assistance to find their own responses. Each community or neighborhood will have its own forms and structures to best nest its particularities within. There are no one-size-fits-all models of solutions. The wildness itself is a tone, an approach. The social service workers must be supported and given permission to show up rigorously, with an alert perception toward detail, and attend to the situation's complexity, not just stick to the forms.

It is a given that some responses will fail; not all will be immediately fantastic. But most solutions are already failing. Even though the dedication and commitment of social service workers are heartbreakingly sincere, they are also burning out. It is not enough to want to help people; caregivers must have the strength to do so. If they are exhausted, underpaid, under-appreciated, and frustrated with the limitations of what kind of care they can give, they may become disheartened and hopeless.

Mutual learning is a form of improvisation that is one of the most crucial skills for the future. The habit of reaching for a manual or a method obscures this most precious possibility zone. People are part of nature, and nature is inherently creative. Nature finds a way. People will also find a way to do so, especially with a bit of support and encouragement. I have seen communities in abject despair find all sorts of unpredicted responses to needs they could not get government assistance for. They did this by improvising together. They met situations outside of normal channels and sourced from what they had on hand, no matter how sparse.

Suppose it were to become part of how institutional and organizational services were offered; providing support to improvise and generate communities that are not strangers to mutual learning is a form of social resilience that is necessary. The solutions will come from the complexity; therefore, the community and family must not be underestimated in their ability to perceive their own complexity. Especially if they have had a taste of learning together when they are not in crisis—the readiness to engage in another, more collaborative way can be accessed. The question is not how to make a better system but how to support people to learn together to meet crises.

As communities and families learn how to nourish life together, there will be movement in how they approach day-to-day life. It is not a project to help individuals produce better incomes or "outcomes;" it is the project of the families and communities learning to learn together. What they come up with as solutions is much less important than the beginning of trying new things together. Even if the projects fail, the limits of the perception of dependency on individual success will broaden. The community or family that knows the feeling of learning something together and possibly doing something they have never done together will be most ready to discover the creative possibilities awaiting in times of crisis. These are the 2nd-order benefits, not to find the solution but to know that we can learn together, improvise together—and bring in many contexts and ways of knowing that loosen the double binds.

To meet the double bind is to reach into another context and try new things together, with another approach, in another tone, and toward another idea of what is possible. It will not be the tone of "the experts will solve this problem," nor the tone of "this problem is the fault of . . ." Rather, it is in the tone of "no one can do this for us, let's figure it out." Groups that can improvise together will be the safest, offer the most healing of collective trauma, and be able to meet crises without turning against one another.

Not surprisingly, some national defense agencies are beginning to focus on neighborly, communal social resilience to strengthen the relationships needed in times of trouble. This opportunity is entirely contingent upon what is possible in the communication. The structure of the existing care systems, premised in industrial logic, runs through our children's schools, medical facilities, economic systems, and even through technological algorithms that categorize us by our interests and online shopping. The procedural processes and aesthetics of the systems are interlocked into an inability to receive and respond to contextual information (warm data). At present, as is evidenced by inadequate governmental emergency responses, it is difficult to prepare for unforeseen crises. Therefore, better preparedness lies in the readiness to meet any situation with an attitude and aptitude for mutual learning. By contrast, the rigidity of protocol undermines communities' inherent abilities to meet polycrisis events together and move through their double binds.

At present, the role of the institutional response system, however benevolent it is in intention, is not feeding the possibility of people communicating in new ways. It is not the mandate of any government or industrial health bureaucracies department to do so. Whether people face the traumatic double binds of war, economic collapse, pandemics, climate crises, or a polycrisis combination of all at once—they will need to be able to help one another on the ground, in their families and communities. I would argue that it is critical to help shift those blockages to communing in ways that allow for and encourage mutual learning in specific locales, with the specificity of the people there. Let them practice; let them learn skills of how to care for each other. Let us improvise, not randomly or in frivolity, but with the depth of our knowing and the potential of our learning.

As I wrote at the beginning of this book—

There will be no community without first communing.

Fig. 50. Leslie Thulin. (2023). *Dawn*. [digital art].

SLOW TRUTH

> "That things 'just go on' is the catastrophe."
> ~ Walter Benjamin

This is the dystopia.

Its counterpart is the vitality of caring and tending to each relationship in each moment ... in the supermarket, in the living room, in the social media space ...

It is also a small hint of a path. It is not what is said or unsaid but the relationships in which the saying is taking place that give the meaning. When people get sick, other people care for them, and when you get sick, someone cares for you.

This is deep code.

Does the shared vitality of life itself matter?

Not because it is economic or political but because it is the soil upon which the future grows. Yes, it matters ... more than ever.

And perhaps, in this moment, some of the shine of the other distractions will fall away. Perhaps in slowing down the pace of life, there will be an opening for more earthbound senses to receive the information that is here.

The truth is slow.

COMBINING

>   And so we carry in our cells
> the stories that no one wants to hear.

I Want You to Want Me to Want You

> "Life is complex in its expression, involving more than percipience, namely desire, emotion, will, and feeling." ~ Alfred North Whitehead

I enter this conversation with the voices of generations that have come before and generations yet to come. They are screaming through the keyhole of this potentially transformational moment in history. Some tell stories of epic, passionate encounters, both found and unfound. Some have rage written in scar tissue. Many agonize in regret for brutally taking that which was not offered. Some meant well but failed. Not everyone meant well. Pain makes more pain, which makes numbness that makes more pain.

Unspoken assumptions about perceived dominance (regarding gender, status, and appeal) inform affections. These presuppositions are marinated into every interaction, sometimes as a rebellion against the past, other times in conformity. The hints and signals of old patterns can be blatant and easily recognizable. Or they can be subtle, making it difficult to find a way out of the deep scripts of past generations. Calling out these obsolete patterns is helpful, but it is unwise to assume that it will banish the patterns so easily. More likely, they will morph and twist into new frames. In the arc of the story of women, bras are an example: Women burned their bras in the 1960s as a statement of emancipation, but now bras have become a symbol of confident sexuality in super-sexy lingerie. My daughter asked, "Why burn bras? They are so pretty?" What can I say?

Imperceptible lapses of perception lurk everywhere. There is no way to know what to do concretely. Wide-angled attention is needed. To be lost now is a good thing; it's a commendable attribute in a potential partner to tread carefully into the territory of intimacy. Rather than cavalier, savvy, and cool—the thing to look for now is someone who is "OK" not knowing what to say or do to encourage the relationship toward physical intimacy. In that caution is a message that it matters that we do not hurt each other: a message that says, "I want you to want me to want you."

The urgent need to stop exploitation is a matter of protecting vulnerable people; new rules, policies, and punitive measures have met it. Criteria are drawn and redrawn in an attempt to effectively refine the guidelines for behavior in the office, at home, and in dating relations to minimize harassment. While these measures are necessary, the larger cultural appetites and sensitivity to harming others must also be addressed. What is the pattern that connects issues of gender equality to the health of the oceans, the crisis of economic inequality, and the rising diabetes epidemic? At first, this may appear to be a string of unconnected, siloed emergencies—until a pattern of repeated epistemological framing comes into view. As children, we become habituated to modes of objectification, fragmentation, and separation, and these lessons get reaffirmed throughout our lives as a way of making sense of the world and removing complexity. Institutions share the pattern of separating departments in education, health, economy, politics, etc. Supposedly, those separations create clarity and order, but what is the cost? As the surrounding world is defined through objectification, what insensitivities and numbnesses are learned? Is this reductionism a root cause of the exploitation that spans relations from gender to ecology?

We live in a world that celebrates "takers." We call it ambition, leadership, and victory. The gentle and the careful get trampled, while the aggressive rise to the top. Takers take. Now, exploitation in all its forms is on trial. The entire ecology, including all but a few economically privileged humans, is disenfranchised. Our bodies have been taken without our permission. I would argue that the survival of our species and thousands of others is hinged to violence that stems from the same habits of objectification as rape and abuse. Objectify—remove the person, family, species, ocean, or any other living system from its contexts, relationships, history, and interdependencies—and exploitation, humiliation, and destruction are not far behind. Cut from the relations that give life, dangling and isolated from the ties to other people, other families, other life—the illusion of it being possible to objectify gives rise to those actions that become systemic, multi-consequences that reach into the next generations and beyond.

The stories that our bodies hold are fluid. It is hard to know what history informs present decisions and how much of the sense of "I" is actually able to act outside of the currents of culture. The cells in my body that know the pain of abuse reach back to my mother, her mother, her mother before her, and countless generations. It is impossible to draw a firm line between what was and will be, yet that is precisely the mandate of this era. A new sense-making is necessary to address the horrors of exploitation in any direction. Acting in new ways, using new language, and making new observations are needed—a process of altering and infusing new stories as they enter our bloodstreams, catalyzing shifts in the alchemy of our relationships, and the old and the new will season each other. It is a dangerous time to be loud and an even more dangerous time to be silent. Either way, the past and future are with us.

Life is complex. It just is.

To be alive is to be in constant interaction with the complexity that exists all around us and within each living organism—and more, as complexity includes ideas, cultures, and the curious, infinite responses between them. For any organism, the pursuit of making sense of life requires participating in the interrelational characteristics of complexity in every moment. The ability to make sense of the surrounding world is fundamental to survival and sometimes results in evolution. To walk across a room, plant a garden, or make love is to engage, mostly unconsciously, with what seems to be an impossible array of complex processes. Sense-making is a process of processes—merging conceptual, physical, sensorial, biological, and linguistic systems within larger contexts of culture and the biosphere. Interdependencies form and inform these contexts. Complexity is vast.

The global crises currently stem from the exploitation of the environment and human rights violations are clear examples of situations too interrelated to be solved by any nation or within any single sector. But complexity is also intimate. Less obvious but equally important is the study of the parallel complexity that exists at the personal level. Interdependencies form and inform these contexts. Decisions made by global leaders and decisions made in the intimacy of our personal lives require gathering information from which to make choices. What information has been considered, explicitly or implicitly? Could a better understanding of complexity at an intensely personal level allow for a more immediate, embodied understanding of complexity on larger societal and ecological levels—and vice versa?

To begin to consider complexity at an intimate level, let us imagine the complex contexts that sex brings together. Sex is sensual, emotional, biological, political, economic, and cultural. . . . Sex is complex. And what about consent?

Reducing the complexity of sex to a collection of rules and procedures is a disastrous way to try to make any sense of it. The opposite of complexity is not simplicity—it is reductionism. The tragedies of sexual abuse are critical socio-cultural wounds that must be faced and addressed. But, to do so is to notice how sexual abuse is framed within the society and culture that creates it. Reductionism as a form of description requires reducing complex "whole" ideas and phenomena to their simple decontextualized "parts." Doing so turns off any perception of vital relational information, resulting in spinning-off consequences of consequences that disrupt, damage, and destroy in unpredictable ways—which is as true at the level of global politics as at the level of intimate sexual consent.

As an aside, short-circuiting complexity is never a good idea. It makes life complicated. Complicated and complex are not the same thing. Complex looks like an ocean, whole and alive with vitality generated through interrelational, interdependent processes. Complicated are what happens when you break those relationships into parts and try to control them, like pesticides on our food and the medical and ecological consequences of the consequences that pesticides have created.

It takes complexity to meet complexity

If we are looking for quick-fix answers and binary memes, we will find them, and they will not suffice to build new ways of life. But, if we can begin to recognize the complexity in our own identities, we may acknowledge that of others and thereby humbly enter another level of mutual respect.

Physical attraction and the body's pull to be close to another person is a simple phenomenon of life. Simple. That physical attraction is an expected experience that most adult humans have known. But, it is also complex. This mutual yearning emerges through combined complex systems from culture to cognition and from the microbiome to material wealth. While attraction includes multiple, inexplicable ways of receiving and responding to the presence of another person, respectfully acting on that attraction requires consent.

What is mutual yearning?

Combining

What, then, are the processes that inform sexual consent? At first glance, consent can be seen as a mutual decision to share sexual interaction. But with consent comes the complexity of all the institutions of our world, alongside all of biology. From law to pregnancy, disease, religion, or finance, consent is loaded with mixed layers of complexity that must be communicated. It is hard to imagine a more abundant source of trauma, destruction, and pain than that stemming from reducing sexual consent to a simple "yes" or "no." The confusion of sexual consent is one of many emergencies arising from patterns of reductionist perception. The global crises of this era all require that more, not less, contextual information be included. Increasing the ability to perceive and live into an ongoing interrelational, contextual understanding of our intimate personal lives is vital to perceiving the complexity of our more extensive, shared survival within the biosphere.

Sexual consent is often described as the binary of saying either "yes" or "no," but it has become clear that this binary leaves everyone vulnerable. Misunderstandings, accusations, pain, and trauma spin out of an approach that reduces such a multilayered and complex interaction to a simple "yes" or "no." Richer communication and understanding are necessary. The context of the relationship often has more relevance to this permission than what is actually verbalized. While the "yes" and "no" are undoubtedly important, it is also imperative to bring greater visibility to the contexts in which consent is determined. While the context of consent appears to be located in the privacy of the relationship between two (or more) people, in fact, multiple contexts reach out into the realms of family, community, culture, and legal and economic status. A reductionist approach to consent excludes these contextual factors, which may even include the survival of those giving consent. Thus, sexual interaction can become a confusing and often destructive transaction that the institutions of law cannot clarify, medicine cannot cure, and cannot be pulled out of the culture(s) in which it occurs to be redefined.

The reductionism with which consent is often described can make it impossible for us to perceive the information we need to make sense of each other's verbal and nonverbal communication. Reducing consent to a toggle between "yes" and "no" does not leave room for considering other influential conditions and contexts. The binary short-circuits the communication, sending incoherent messages into the relationship.

Given the complexities within a sexual encounter and the culturally habituated mechanistic presumptions about controlling unpredictable experiences, it is unsurprising that there has been a demand for a way to lock down communication around consent. By eliminating the variables surrounding consent, a well-intended attempt has been made to provide a straightforward way to communicate in an unclear moment. The choice of "yes" and "no," while offering a clean line to define consent, has brought subsequent abuses and horrors. "Yes" does not necessarily mean yes. "No" is not always no. My mother, born in 1929, is from a generation that considered it unseemly for a woman to say "yes." If she was too eager, this sullied her reputation; it indicated that she was easy and a turn-off to the men who like the chase. The no-means-yes flirtation was there to help keep her dignity intact, which was prevalent until the sexual revolution for her generation but is still true today in many communities and cultures—no doubt this generated and still generates confusion.

On the other hand, people sometimes consent when they have no choice, when they feel afraid not to, or when they are bribed or manipulated. Likewise, people say "no" when forced to by law, culture, or religion. Additionally, someone may think at first that they want to be touched, and then if something does not feel right, they may want to stop. Some people guilty of abusive sex are never accused . . . and should be; others are falsely accused. The imposition of a binary does not produce the contextual information needed.

By saying that consent is complex, I am NOT implying it is unnecessary. Victims are not culpable. Not at all. It is that consent is more than "yes" and "no" and that without some deeper understanding of the contexts of consent, justifications will be made that hurt everyone. Additionally, the idea that we can use reductionistic approaches within the complexity of consent leaves a vast horizon of loopholes that can and have been used against each other. The binary may be expanded with explanation in the courtroom or home; however, the attention toward a binary reduces our ability to understand the vastness of our interactions—it narrows our approach to perceiving and understanding the complexity. Our interactions turn towards narrowed stories and statistics that cannot hold the water of the life we live within.

How many people have had sex with their partners when they did not want to because they felt that they needed to keep them satisfied to keep the family together? Is that consent? Well, legally, yes. But, in the particular reality of those people in that bedroom on that day, it may be more complex than a simple yes or no. The question of consent seeps into economic survival, a murky brew of culture and history lingering in explicit, non-verbal, and non-conscious ways. I am not sure how to know when yes is yes and no is no. Without consideration of context, the differences between mutual desire and transaction are blurred.

Transcontextual Considerations of Consent

A binary or polarity will remain when any complex circumstance is short-circuited. When the circumstance is described in reductionist terms, its relational messiness can appear to be washed clean; the outcome will appear as a clean linear solution with a toggle switch. Good and bad, left and right, us and them . . . but the choice is a trap, a lure, a false flag. The complexity is not vanquished just because it has been removed; it will return. It is still there and will continue to be there underneath, feeding possibilities of further destruction, hurt, and suffering. And what will we learn? What will a perpetrator learn? And what will a victim learn? Where will this leave us? Leaving these seemingly clear solutions built through the court of law allows more precedent to be built upon the obscurities. And we continue in this fashion while we continue to ask, "How did we get here and why?"

It is easy to shrug off the possibility of ever being able to take into consideration the many contexts of contexts that reach out from the minute detail into the cosmic eternity. But it can, and it must become habit to at least approach the complexity with the humility of knowing we cannot know it all. From there is the possibility of including as much relational information as possible, personally, professionally, and publicly.

COMBINING

The following is a cursory and incomplete listing of some Warm Data contexts that inform consent. It offers a glimpse into how these contexts entangle, magnifying the effects of reductionism. Consent is not just about sex. The contexts listed are not the only contexts that inform consent, nor are they exhaustive in their transcontextual description. The intent is to provide a small taste of how multiple aspects of life are interdependent in the process of considering "consent." The warm data arises as the contexts start to reiterate, overlap, and link to the reader's personal experience. A budding awareness of how consent ties into all aspects of life reveals the liminal realm—between contexts where substantive change is possible.

OUR IDENTITIES AND EDGES WITHIN CONTEXTS OF CONTEXTS

THE SELF: The self is complex. Engaging in a conversation with another person is a coming together of ecologies. It is only possible to consider consent by first developing a notion of self. Who are we? Many contextual processes make up identity. I am a mother, a daughter, a friend, a professional filmmaker, a writer. I am American, and I am European. I am a traveler. I am the forty trillion organisms living on and in my body relating to the larger ecology. I am language and gender. I am my tax records. I am an ecology of selves. These selves may not agree or seek the same kinds of relationships. And where is the edge of the self? Is it the skin? Or is it the ideas that are relevant to that person? Is culture the edge of the self?

Reducing identity to a single context is a common way to claim belonging in a particular group. But, to singularize the self in this way is a harsh edit of so much of who a person is, was, and will be. Expanding our notion of self offers insight into the variables that make us who we are from one day to the next—ever calibrating, ever responding to the world around us. Familiarity with our own complexity is a step toward the generosity of giving others permission to be complex, too.

ECONOMIC: What are the economic repercussions of denying consent? Often, consent is given by the necessity of personal economics. The loss of income or position can be an unspoken casualty of saying "no." Safety and security for oneself and children are at risk if a relationship providing economic support is withdrawn. Survival can depend upon sexual consent. Differentials in pay, sexual attractiveness, and economic success are tightly interwoven. Economic imbalance in the relationship becomes a point of leverage that affects consent. Dominance, status, and influence are often gained through material wealth. Coupling into wealth has a long history of manipulative leveraging on all sides.

EDUCATIONAL: Gender roles and expected behaviors are woven into the classroom experience. From how history is written to how literature depicts heroes, leaders, and even the study of feminism, younger generations learn that the world they are entering is different for men than for women—the interpretations and learnings around consent are affected by this and young people learn who has a say and who does not.

POLITICAL: The law cannot effectively meet the challenges of consent. Rape and abuse accusations require evidence, which is often impossible to provide—leaving the justice process dependent upon "your word versus mine." Both victims and those falsely accused are unsupported by this inability of the law to take the situation's complexity into account. Meanwhile, many abusers use loopholes in the law to escape punishment. The law is dependent upon a binary that is too often irrelevant and insufficient to the case. The circumstances around consent can be so complex and overlap so much that it is impossible to reach a definitive statement of "legal" or "illegal." Contextual information is necessary. "No" is clearer than "yes." At some point, most people have conveyed "yes" when they did not really want to—some for more urgent reasons than others. When factors such as someone's housing, food for their children, continued employment, or religious status are leveraged against intimacy (even indirectly), it is likely that "yes" is more complicated than it sounds. Laws of marriage and the distribution of wealth and property all figure into the overall set of political ramifications of consent in nearly all cultures.

The language of rights is mismatched to a reality of interdependency and vulnerability, but also to the idea of intimacy, especially sexual intimacy, as something freely given. The word "consent" holds a meta-message: an agreement, a contract, a signing away of rights. The principle of consent is that no one has a right to another person's intimacy. Claiming otherwise is an entitlement, a vast cultural pattern we are trying to escape. Yet everyone is tied to others by histories of relationship, reciprocity, succoring, empathy, kindness, cruelty, distrust, hate, emotion, and interest . . . these all affect consent in the moment and the wider context. It is also true that attraction is affected by social factors and bigotries; the unpopularity of some groups or categories can make intimacy harder for many people, and it is not an entitlement for their members to discuss how intimacy is not an even playing field. But, this right to observe and discuss how the world affects you does not extend to advocating the right to leverage or guilt-trip others into intimacy. The sovereignty of the right to consent or not is key to the autonomy and dignity of every human being.

Historical:
"Go for it" ~ The motto of the modern world.
"Just do it" ~ Nike slogan

Having met the monster that is the great-great-grandchild of colonial history, we have a severe re-direct to address. The idea of prototyping a change model to stamp across the globe or the idea that we should "go get" the change we want is steeped in the same thinking patterns that have created the destructive violation of the world. The thinking patterns of "taking," "having," "owning," "leading," "controlling," and "winning" link competition, capitalism, colonialism, and numbness.

Even though people knew their aggressive actions were hurtful and illegal, they were accepted in the culture. "Locker room talk" and the related idea that "boys will be boys" has been and often still is considered "normal" and even "natural" behavior in a way that provided implicit acceptance of exploitations. The idea that if you want something, you should "go for it" is an idea that carries with it a history of triumphant conquests, possibly precisely because careful permission was not sought. The idea of "just do it" is the backbone of industrial "progress" and leaves a wake of ecological, economic, and social destruction. The same attitude in sexual advances is accepted as assertiveness and, until now, has not been questioned or revealed as predatory.

Sensitivity to exploitation as it is occurring around us at all times is challenging. Numbness makes it easier to continue. It was not so long ago that women were considered property. In the context of sexual interaction, a new language is forming. Still, other numbnesses remain. The human rights violations and ecological destruction that produce food, clothing, and technology in our daily lives are still largely unseen. These awakenings are necessarily linked. The treachery of greed, which has generated appetites for taking from the bodies of others, is the same treachery that generates appetites for money, fame, and status. The appetite is numb to the vulnerability of living interdependency. It has to be numb if it is to succeed in such a violent world.

We cannot change our trajectory now using the same sensibilities that got us here. The sensitive get trampled, while the aggressor, the oppressor, and the "assertive" succeed. This is a noise so loud that the new growth cannot be heard, but receiving is so different from taking. Isn't it so? We can choose to emulate the kind of people who enter a room without assessing what position to take, without considering what status to take. . . . or we can recognize that new sense-making means receiving new information and staying open to the context. It is all about sensitizing into what nuance and micro nourishing is necessary to give.

Cultural: What does it mean to be desirable? Where do these ideas come from? Notions of femininity and masculinity are generated in the culture. Marriage, partnership, and sexual "norms" exist in cultural expression. Language, fashion, film, music, art—all carry the messages of what consents are expected. The "yes" and "no" binary is wholly framed inside a culture. The culture may have a law that says sex without consent is rape. But if people cannot say "no," non-consensual sex is normalized. Cultural pressures to keep marriages intact or to be a good spouse may require fulfilling the needs and desires of one's partner. To say "no" would be to break the partnership; to say "yes" is to consent to sexual contact without wanting to. The stories that inform notions of coupling, shame, goodness, and well-being are all infused with culturally shared assumptions.

Media: The media provides the conduit through which many unspoken assumptions are conveyed. Consent is woven into ideas of dating and love stories. But these assumptions are not just directly described through the imagery of love; they are implicit in other ways. News, science, or opinion pieces that often have nothing to do directly with sex are infused with sexuality—giving silent, hidden messages about consent.

The stories of what love is, what sex is, what happiness is, and so on spin out of the media. In the most public ways, these images are of the heteronormative, happy (usually white), thin, and muscular twenty-something couple at the beach in a state of constant mutual eroticized desire. In that image, the rays of disenfranchisement are myriad. The "perfect" couple is young, making older generations invisible sexually. The "perfect" couple has a high economic status, and is white and skinny, which leaves out most of the world. The "perfect" couple is heterosexual and interacts along hetero scripts of affection. Boy initiates, girl accepts. But what about the infinitudes of other forms of interaction, gender possibility, and the expression of more unscripted sexual pleasures? Are they bad? Should consent be contingent upon the portrayal of a certain image?

FAMILY: Each family has its own culture and has scars of its own. I remember the day my mother told me the story of when she was raped. I remember telling my daughter the story of when I was raped. The generations are linked in their trauma and in their healing. Somehow sharing this story was an instruction on how to survive in the world. Family is the matrix through which all the contexts of our lives intersect.

Sexual consent can become twisted up into a muddle of the hopes of keeping the family together, protecting children, protecting the home, and protecting reputation. How many marriages or domestic partnerships have sexual encounters that are not wanted but believed to be obligatory? Not all nations today consider non-consensual sex within a marriage to be rape. Religion and culture are relevant to a family as well. Sometimes there is nothing more pleasing than to please the person one loves, even if one's desire is not boiling. But to do so in fear of economic consequences can lead to resentment that can become a lifetime of marital date-rape. Does that protect the family? Is there support within the marriage to refuse each other? Rejection never sits well. But false consent is also toxic.

ECOLOGY: Attention to destructive patterns of exploitation between people may help make the exploitations between humanity and the environment visible. The violence of reductionism lies in the objectification that comes from separating living systems into parts, cut off from their relational vitality. Complexity is concealed when severed from interrelational contexts, limiting understanding, respect, and care for the interdependency of life at all levels. Objectification leads to exploitation. The description of human beings as objects, for example, using the metaphor of a "role" or "resource," is a dangerous reductionism, leaving us all vulnerable.

The decontextualizing of human beings from their bodies, communities, and cultures devolves toward ownership and economic valuation. Things are owned, not relational processes. Isolation from context is premised upon machine metaphors in which parts can be broken, replaced, or fixed as components. Mental, physical, and emotional health are interdependent processes that reach well beyond the identified parties and into families, communities, and future relationships. This ecological pattern of interdependency is a door to recognizing that understanding the complexity of the self in intimate relationships is a necessary contribution to understanding complexity. Perception of complexity in an intimate way extends to care of that same complexity in other relationships, such as to the biosphere.

TECHNOLOGY: The mechanistic efficiency of technology seems only to increase. The "next-gen" stuff is always faster, smoother, and trickier than the previous model. We are told that machines can make sense faster than we can. I would ask, what kind of sense? Programs for sorting and categorizing analytic data and statistics are creeping increasingly into our most private information, but do they know us? Sci-fi stories have explored whether computers/robots can experience sensations and feelings for decades. And when we make sense of the world through our devices, how does that sense-making differ from the matrix of interlocked messages met by brain and blood in conversation, in culture, consciously and unconsciously, explicitly and implicitly?

The culture of dating through digital apps has fundamentally altered the scripts of romance in ways we are not yet aware of. I feel that shopping online for love has already become addictive for some. It can be a continuous hunt for the next desirable contact. But there is no denying the experience of people who find each other and find love through these very digital technologies. Is it so different from meeting somewhere else, like a park, a bar, or a friend's wedding? This speaks volumes about the placement of concern around consent as a legal issue. It also speaks to the misplacement of attention to consent as though it were a "thing" that could be separated from the complex contextual joining of two (or more) people.

We do not know yet what the next generations, whose free moments are spent with technology, are learning or not learning in the realm of non-verbal communication. Are they adopting an instinct of linking and connecting from their time online? How are their bodies developing differently in lieu of hours spent in physical contact with soil and grass and climbing trees? The subtleties and nuances of reading another person face to face are somehow affected by emoji culture, but we do not know how yet. Sexual development in the era of online pornography is not comparable to the old days of stolen copies of an uncle's Playboy magazine. What is happening to the relationships between body and imagination, and how that will play out in sexual encounters, is not known yet—let alone the way it will alter mutual sexual respect.

THE CONTEXTS ARE CHANGING

Embracing the complexity of the transcontextual processes that every encounter includes is a beginning. Still, the project of making sense of our own and other people's communication about sexuality will always be baffling. Inexplicable physical and emotional responses defy comprehension as a single direct signal. Yet, to achieve clarity about the encounter, such a singularity is precisely what both parties are told to seek. In the unfathomable genesis of hormones, fantasies, and complicated emotions, sex transports us out of the direct zones of plain language. The complexity of sexual interaction requires recognition of and inquiry into the contexts in which the interaction is taking place. Habits of making sense in terms of separation, silos, and singular causality obscure the complexity of consent. Subsequently, respect for the complexity is also obscured. The context in which the communication is taking place limits what is possible to say.

The topic of consent is filled with instabilities brought into public view through the #MeToo movement. With this collective attention to the mishaps and abuses of consent, there was a loss of "sureness" about discerning mutual consent or even discussing it without triggering old, inadequate language and oversights; this is how it is now. The context in which this conversation about consent occurs has become rapidly blurred.

Confusion about defining acceptable conduct is coupled with the fire of fueled fury over a world that has let us all down. If advances are made now, they are read through a different set of contextual lenses than the same advances might have been read even a year ago. Despite the long history leading up to the viral #MeToo moment, it happened swiftly when the shift began. As a generalized example: Now, if someone is persistently making unwanted advances, the assumption is that the person is making them despite knowing what damage is caused by that behavior. Not long ago, that person making the same unpleasant advances would likely have been described as clumsy, old-fashioned, and pushy. Now, millions of voices have spoken about the violation, damage, and torment of that behavior. Now, anyone who behaves in this manner is viewed either as contending with some sort of pathology or as a singular asshole. In any case, now, they risk losing status financially, legally, professionally, and socially. The same messaging has a very different meaning than it used to. The context is changing.

Our hearts were broken by The endless stories of sexual abuse witnessed with the #MeToo movement brokes our hearts. And we must admit that we do not have a clear rule book. There is no standard. Some people cannot be told "no" strongly enough, and others seem to be at ease reading the signals of mutual desire or the lack thereof. For all the invasions of my privacy I have experienced, there have also been people who honored the communication, verbal and non-verbal.

The world we live in is allergic to complexity. It will not allow the gray area (or complexity) into the room. For those who cause harm, there is often only a choice to gaslight the survivor and deny that anything happened or plead guilty, which will land you in prison and likely compound all of the trauma that might have caused you to harm someone in the first place. There's no place in a courtroom to say, "I didn't realize how harmful this action was, I didn't mean to hurt you, I don't want to hurt people, etc." When there is talk about accountability, there is rarely acknowledgment of the complexity of what it might entail to hold that accountability in a system that doesn't have any way of valuing whether someone has changed or has gained the capacity to perceive the harm they have caused in new ways.

My experience with the legal system left me realizing that even if a conviction was made in my rape case (which did not happen), my body would still carry the memories and the terror. I carried the name and address of the person who kidnapped me into the police station. Three men on one side of the table, me on the other. They sat there and only said: "It's your word against his."

Nothing makes much sense, yet sensing is critical to finding a new language of consent, new forms, and new gestures; this process runs across the contexts of our lives. Society is rife with rape and sexual abuse, pulling against well-versed discourses of feminism. Reaching far beyond the back of the bar or the bedroom, the contexts that inform consent are inter-systemic. Identifiable patterns of exploitation and oppression bind personal daily habits and expenditures to the suffering of others and the ecology. I live in a pull of contradictions.

Living in a world of exploitation becomes normal, feeding the distrust and the reasons for distrust between people. I am unable to remove the messages of my culture from my body. The perplexing incongruities of society purport to care for citizens but put business first, entangling justice with profit. I carry these paradoxical perceptions. The mixed messaging of assumptions around sexuality is informed both physically and intellectually with centuries of pain. It is marinated in our language, our bodies, and our emotional responses. Constructs of masculinity and femininity upheld by the structures of historical roles of cis men and women are disappearing. The familiar walls are crumbling. They were weight-bearing walls; they held the architecture of sexual expectations.

A clean start with a new set of rules is impossible. Wherever the future leads us, we take our existing scars. With that in mind, I am curious about what a "healthy" sexual relationship is for individuals and the community. I am trying to get out, get in, get over, get through . . . get there: get to where the complexity of sexual interchange, interaction, and intercourse is given room to shift.

Lost Together, Learning Together

As a student of complexity and systems theory, I have found that it is often the case in inquiry about larger contextual processes of a system that one's loyalty comes into question. It makes people suspicious when they cannot discern whose side I am on. Let me be clear: My loyalty in this exploration is to those brave people who find themselves enamored with another person in a world that is changing shape, and not knowing what to say or do, they pause. They are pausing because it matters to them that they form relationships that do not damage those to whom they wish to offer affection.

I believe that establishing future patterns of expression of affection lies with these people who, in this moment, are attentive, nervous, and careful. There is no measure of the regret I would feel if my attempts toward finding possibilities for understanding one another were misunderstood. I am not here to betray, vilify, or victimize. This articulation is perilous, and I may fail.

The confusion of this era contributes to more confusion, even when it comes to what we are confused about. I suppose that those who do not care about the damage they may inflict upon others still do not care. But, those who care to assure mutual consent are stumbling between old and new mistakes.

The old scripts of flirtation and courtship have thankfully begun to rot; their hierarchies and exploitations have been revealed. But, now what? A necessary lostness has come with the shakedown of so many who abused their positions. Now, especially since the #MeToo movement began, caution is the emerging script for some, while others indulge in nostalgia for the clarity of previous understanding of roles and responsibilities. There are sharp consequences everywhere on the topic of consent. The slightest mis-saying will scream louder than any attempt to unsay it.

This moment of confusion is a portal to the next phase of our ways of being together. Perhaps there is another territory of contact and communication. It will not be a prepackaged solution or a set of steps to a promised harmony. Fractures forming in the perceptions of gender and consent are opening the possibility of seeing sexual interaction in new ways.

Desire

Deep in the complexity of consent is the needed existence of mutual desire—a mysterious territory. What makes people attracted to each other is one of the most inexplicable, unknowable phenomena of human interaction. Still, even though the sources and causations of sexual attraction are beyond description and predictability, it is imperative that the complexity of desire is considered. Especially, hopefully, mutual desire.

> We have these earthly bodies. We don't know what they want. Half the time, we pretend they are under our mental thumb, but that is the illusion of the healthy and protected. Of sedate lovers. For the body has emotions it conceives and carries through without anyone or anything else. Love is one of these, I guess. Going back to something very old knit into the brain as we were growing. Hopeless. Scorching. Ordinary. (Erdich, 1998, pp. 147–148)

Sexual appetites form and shift in territories of the self that are not necessarily rational, verbal, or explicit. People often do not know why they want what they want. They cannot explain or justify their stirrings. Every great love story and poet has explored this universe of a charming and dangerous mystery. How one person perceives another involves vast processes of hidden information. The memory of pleasurable and painful experiences, cultural messages of beauty and strength, and ideas of happiness, fulfillment, and success—all lurk beneath the realm of conscious verbal analysis.

Nothing makes much sense in the murky physical, emotional, and cultural connections that take place to produce attraction. Ironically, sensual processes make their own sense, with little regard for the way sense-making might occur in rational, logical terms. Contradictions, rebellions, and forbidden-ness are so prevalent in this unknowable realm of behavior that they are clichés in most love stories, fiction or not. The contexts of body, intellect, history, and so much more combine indescribably. Tread carefully on any path of ideas or wordings that touch this wound. How do people know if the consent is mutual? While it might seem advisable, an easy rulebook of instructions will only result in more confusion. The complexity will continue to elude predictability, wreaking havoc until an understanding of the contextual processes is given its due. I am raw. I am nervous as I write this. Any small misstep may trigger venomous ire from all sides. Silence tightens.

Now what?

"I want you to want me," the title lyrics from a 1970s song by the band Cheap Trick. This reciprocity is not enough—it becomes a game of luring, attracting, and seduction that is too easily imbued with dominance, power games, and abuse. A deeper consideration of consent has another dimension in it, which is where possibility and complexity reside.

I want you to want me to want you.
A is attracted to B (and C and D and E)—there may be more people in the relationship too.

A pays close attention to B's verbal and nonverbal communication. Is there perceived reciprocity? So far, it is, "I want you to want me," and the next level is crucial in this new era. Is A turned on by B being turned on by A's attraction, and vice versa? Expanding this also into polyamory and desire that includes more than two people.

Let me break that down. A wants B (and possibly multiple partners) to notice and even be turned on by the desire of A. But more than that, A wants B to want A to feel that desire. It's not just "consent" that matters; it's that potential partners are desirous, craving, and excited to be wanted by the other—who likes it, too. Here desire, which can be so dangerous, so greedy, and so unthinking when it is felt by one person in isolation, is pulled into looping reciprocity and becomes mutual. The horizon changes shape here—completely. The contexts of consent become important to perception for both people, the community, and even the larger ecology, marking a change in approach.

We have no idea how to do this. We have no stories to show us. We have no elders to describe it. We simply must walk together into this darkness and keep trying. This lostness is not a cop-out. To reside in ambiguity requires rigor and integrity. Not having a script or a protocol is hard work and takes much more attention to the warm data. Looking for the complexity in another person and responding to it is a process of care and curiosity in the minutiae of each moment. The responsibility to change our systemic cultural pathology around sexual abuse resides in each of us every moment of every day. We are all carrying the scars of our mothers and fathers. Likewise, we are all contributing to the conditions in which the next generation will make sense of these things. We are damaged. Somehow, through our damage and our inability to perceive, we have to find new ways to respect each other, to enjoy each other, to find and give sweetness, love, and passion to one another.

I think it is the trying that matters most of all. To care that someone you want wants you to want them each time you come together intimately is a recognition of their complexity and your own—which means finding new ways of reading gestures carefully. Speaking in a language that does not pull old scripts into the new conversations will be necessary. We will need to practice listening with the body into unfamiliar sensations of attraction and respect as they unfold in new contexts of culture. Playfully together, we may find possibilities as yet unimagined. For the coming decades, we will be holding our expression of affection lightly so that when our inexperienced attempts crash, they do not demolish life.

> "If the future is to remain open and free, we need people who can tolerate the unknown, who will not need the support of completely worked out systems or traditional blueprints from the past." (Mead, 1928, p. 348)

> **Addendum to my own healing*
> *I am un-shadowing.*
> *Wind is the song. There is singing.*
> *I am a root reaching into the mud. There is sensation.*
> *I am skipping along paved streets in rain, high-heels, sweaty cheeks. There is a young woman in the city.*
> *Night was the beat. There is the street, the heat, the hips.*
> *I am skin & heat & mind & earth. There she is.*
> *I am visible again, & it is such sunlight. There I am.*
> *My scars are not hidden.*
> *There we are.*

* Previously published in *Journal of Design and Science*

Beyond words ... there is a place we meet.

Fig. 51. Vivien Leung. (2023). *Caring*. [digital art].

Fig. 52. Vivien Leung. (2023). *Transmitting Vibe*. [digital art].

SILENCES

Silence is only sometimes stillness,
Harboring unquiet secrets,
Whisperers wagging their tongues,
A reticent lull,
Or a pause to digest a thought,
A comma,
Reorienting in an unformed sentence,
Hold –
While chasing lost words that refuse to form,
A reserve, bidden by generations past,
Who have learned to withhold,
The same hush is forbidden by others,
Obliged somehow to say everything all the time.

Shhhh.
To be silenced,
And to be silent,
Is not the same lack of noise.
It is not silent,
it is always between,
it is what allows,
it is what stops and goes ...

PREDATORY SKILLS

The story of fallen heroes is reframing the notion of what it is to be relevant.

Through the filter of the hero's story, nature's predators were perceived as valiant, triumphant, and epitomized strength. Individuals were celebrated for their cunning, savvy, and
ruthlessness.

The mythology pivoted around normalizing this.

We were told.
"That's how life is. You need ambition to be a good predator. Get the money, get the award, get the stuff, get respect ... Step over whoever you have to on your way to the top. The powerful are the predators."

And then, it became clear that the survival of our species requires sensitivity to well-being across multiple cultures and species. This is a sensitivity that is unlike anything resembling predatory skills.

Combining

HARVEST

Be careful ... A hasty harvest is an un-feeding.

It is non-food.

Harvests are patient history and quiet roots that start tiny.

While compost is necessary for the growing season, do not eat the compost. Wait. The eggshells and onion skins bear little resemblance to the conversation at the table where the meal they fed feeds you.

Wait ... The industrial craving for instant food, ideas, and plans distracts from entering ecological time.

An apple seed in the ground is not an apple — it requires seasons, years, changes in light, wind, temperature. It must reach up into the spring sun and down into the winter earth. The blossoms need the insects and the song of the baby birds.

This is not the time to harvest; this is the time to let these ideas come to find other contexts in the underground, in the wilderness.

Notice the un-ecology of the question, "What is the point?"

A meadow never asks such a thing.

Life is making life. Nothing less.

The harvest is a time of golden-brown reds and deep orange tones.

The harvest is marking the long journey into the squishy death and rot that offers new compost.

It is the last moon before the cold comes. Do not race into winter before it's due.

Harvest is long in its arrival. Sometimes it comes with an empty basket, leaving its lost leaves and roots to decompose — feeding the possibility of next year.

Likewise, let the ideas and stories of today remain in the undergrowth — let them find new contexts, new contacts, and new life ... let the movement of relating follow its own reachings.

Let it touch and rename itself as it moves. This is the beginning; this is the middle. Let it form and inform.

Do not ask what you will get out of this. This is the making of next year's rain ... the point is the meadow of a hundred years from now.

Ecology is not instant. You cannot run to the shop and buy it.

There is no summation, nothing to put in your pocket, only idea-seeds in warmth and moist history.

Relax ... The possibilities are brewing, stewing. Let them wander and shape in ways you may have never considered.

COMBINING

```
TO LIVE IT

To recognize the field of concepts when it
looks like -
a fever,
an argument,
a carrot,
a glass of wine,
the rain,
language,
sex,
love,
kids ...
```

Ecology of Communication

Every communication is bringing something into an existing ecology of gestures, tones, codes, expectations. What will you bring? What have I contributed to this ecology? Will it grow into lushness, or have I eroded that possibility? Have I made room for new life or poured salt and bleach on a garden of what might have been mutual learning? How has my communication changed the ecology into which you can respond? How can we communicate about ecology if our communication is not itself ecological?

> In fact, the problem of how to transmit our ecological reasoning to those whom we wish to influence in what seems to us to be an ecologically "good" direction is itself an ecological problem. We are not outside the ecology for which we plan—we are always and inevitably a part of it. (G. Bateson, 2000, p. 512)

I grew up in a household in which communication was considered so essential to life itself as to be sacred. To manipulate communication, to justify twisting it, or to violate it was considered a vulgar violence to life itself.

My father learned this lesson the hard way. It nearly killed him. Perhaps the extreme horror he experienced has given me a sense of urgency to attend to intangible processes that steer the domain of communication. In World War II, he was given the job of tampering with radio communications between the Axis countries. He took the job because he was dedicated to the task of stopping fascism, and this was the job the Office of Strategic Services (OSS) gave him. But it came at a price to his soul. The deviousness of the job was its insidious subtlety. He did not blatantly disconnect the communications between the Axis forces; he tweaked their communications just enough to generate distrust and confusion.

He readily said later the duplicity of this task sent him into a bleak despair. By exaggerating the information just a bit, he created a plausible scenario that would make sense to the receiver while generating distrust within enemy ranks. This exercise was distorting communication and destroying the possibilities of communication—the source of interrelational vitality. He was suicidal after doing this work. Even though he was eager to help in the fight against fascism, the way in which his knowledge of pattern and culture was used by the Allies nearly destroyed him.

Combining

By the time I was born, thirty years later, this attention to communication had become more than a theory; it was a way of life. In an effort to limit damage, my father spent his days and decades practicing perceiving complex relational processes through anthropology, information science, ecology, psychology, and art. He could do this better than anyone I have ever known. His communication from the breakfast table to the conference keynote was an endless, curious, caring, and passionate artistry of moving the ecology of communication toward a little more possibility of life to make more relations wherever he could.

Although I did not realize it at the time, I was learning from both his way of attending to communication and his regret at having tampered with it. The regret formed a sadness and a warning that was always with him in the undertones of whatever he was doing. His regret revealed his affection for life which was also there in the undertones of whatever he was doing. He loved communication, and he broke it. This pain never went away. So, I learned early in life that remorse for having manipulated communication is a pain that runs deep. I also learned what it looked like to be careful.

There has been keen interest on the part of the so-called "good guys" in making a new Cambridge Analytica that would swerve public interest toward more sustainability or social justice goals. The thinking behind this is that since it is such a potent tool, it could and should be used for changing hearts and minds toward protecting each other and our planet—producing "this" change in epistemology. Why not? Because it is manipulative and inherently violates the ecology of communication and dignity of relationships. You can't lie your way to integrity. The ecology of the communication is wild and must remain open-ended. The best I know to do is ensure my contributions to the brewing future are careful, thoughtful, and warm.

Symbiosis, the central process that creates life, is organisms living in contact with each other. It follows that dividing, separating, isolating that contact is extinction. You and I abide in many ecologies—we ignore them at our peril. We are symbionts within ecologies of people, our bacteria, our socio-cultural ideas, as well as the earth, plants, and animals. Staying viable in this ecology of ecologies is essential; the alternative is isolation, which is death and obsolescence.

I need the ecology of bacteria that live in and on my body to digest my food and to give me an immune system. I need the ecology of my family and the people close to me; human beings are not meant to be alone; we need emotional, physical, intellectual contact. I need the ecology of socio-cultural ideas and institutions for food, medicine, schools, and so on. None of these ecologies is possible without the larger ecology of the plant and animal life of this planet, which gives peace to the soul, as well as air, water, food, and everything else. All of these ecologies are formed and informed through constant communicating.

A dandelion that grows in an urban garden grows lush, tall leaves, and big blooms. Its root system is small by comparison to the same plant grown in the Alps, where the wind and weather are harsh. The alpine dandelion grows its roots deep to hold on to the hillside in times of stress; it keeps its leaves close and small, and its flowers are miniature. Ecology itself is a conversation in context—sometimes in hidden root systems, sometimes in shameless blossoms. Human beings also learn to keep themselves hidden if the situation is unsafe—we learn to lie and learn not to show affection. We can learn to place our strengths where others won't see them—we learn to stay quiet—be invisible. Or perhaps we learn to catch the attention of others by shining too brightly.

My history includes the odd mismatch where New Age and scientific discourse mixed. In California in the 70s, as a child, I sat at tables with scientists and politicians who spoke of stopping ecological damage and social justice fifty years ago. I was also romping around in self-help retreat centers where my father sometimes gave lectures. These ways of thinking were not in approval of each other. After his death, I spent my teens in the punk rock generation (not the Nazi skinhead sort, but the social justice punk rock.) When 80s rap met punk rock and dub, we thought we had found the mix to break through. Like generations before us, we were ready to change the world. But, like for the generations before us, the world was not ready to change. All of these are responses to a cultural norm which is suffocating communication.

Self-help is rife with shallow methodologies and clichés; punk rock is raw and hopeful that screaming loud enough will rip a hole through oppressive, exploitative systems. Though that was long ago, the rebellion we sought is needed now more than ever. For that matter, the rebellion my father sought is still needed, and the rebellion his father sought is still needed. . . . That is the hillside my dandelion is rooting into.

By contrast, the dark horror of this era is a caustic communicational divisiveness fueled by those same relational needs. Information troll farms, Cambridge Analytica, Aggregate IQ, and others employ people to pit people against each other. They are using context against context, using relationship to break relationship. These entities are a part of daily life now, stirring up culture wars and splitting families. Political polarities are all frothed up by decontextualized information and journalism that sells controversy instead of communion. It is impossible to know what is real, who is trustworthy, and where they got their information.

We are living in a world that is very much like the task my father was given. Communication, which is what holds us together in our ecology of being, has been sabotaged. This water of life that gives communion has been poisoned with fake news about fake news, which was always fake anyway. Thankfully ecologies are not so easy to control, and authoritarian attempts, while tragic and horrifying, do not last. Life gets loose and pops out some new uncanny possibility. Some unintended consequence is always on the way.

The question is, "Will it bring life-giving communication or divide us further?"

Communication has been squeezed into all sorts of contortions since before we know. It did not start with troll farms. The separation of personal life from public behavior has its upside, but the seemingly benign monotone of professionalism has contributed to a whole mess of dehumanized, decontextualized perceptions. Attempts to exclude the irrationality of personal intimacy come at a cost of important tacit information that is unclear but is there nonetheless. Another attempt is the anxious grasp for objectivity which is signaled through language without warmth or relational context. This language justifies horrific atrocities in the name of efficiency and results.

There is a confusion now; no one knows what to believe. I think this is partly because without relational information—warm data—the information has become so decontextualized it is untethered from the contexts in which it made sense. Who paid for the research? What questions are on the survey? Which measurements were left out? Are we in a courtroom or a classroom? Are we on a conference panel or at a poetry slam?

Each context has limits into which the communication will ping—confirming or altering. There are things that cannot be said because the context disallows it. There is no rule, no law—none is needed. The body knows these limits even if there are no words for them. It is those limitations I am interested in. If the communication were taking place in an ecology with different aesthetics, would it be possible to be in relationship in another way? Would that not be the most significant form of systemic change? This work is warm, full of beauty, and it is careful and curious . . . and disorienting. It is necessary to disorient toward new orienting. Where is the communication located? Is it in me or in the conversation with others?

The experience I am having of our conversation is landing within a lifetime of conversations, each of which has contributed to my idea of what a conversation can be, could be, should be, and should not be. Our conversation is not isolated, but rather it is illustrated by our histories which are informing the vocabularies and tones, metabolizing it into the way it will land for each of us. I may be charmed by you, or I may be irritated. You may remind me of someone. What you say is changed by the context of how I receive it. So, what will I say? If I say nothing, what will my silence convey . . . to you? Other people may feel very differently about the same statement or tone of voice—the context matters, the listener matters.

I remember a situation where a grandmother was scolding her small children when she would come to visit. She thought she was doing her duty to help them become well-adjusted citizens. The kids quickly began to dislike and avoid their grandmother. In other situations, they might have accommodated her sharp tones. But her visits were infrequent and short, so they did not have enough time together. There were not enough other forms of communication to dilute the scolding or to keep it contextualized in a relationship of care. To the kids, it felt like the only things she ever said were critical; they did not have a diverse ecology in their communication to hold her sharp words. The ecology of the communication was too acidic to grow the sweet fruit of intergenerational mutual learning. They all missed something important in those years.

If I say "Thank you" to you in many English-speaking countries, it is expected that you will reply with "You're welcome." If you do not say, "You're welcome," another communication has taken place. The lack of the coded response comes to express a sentiment that I am "not welcome." It is an indication that whatever I was thanking you for may have taken place in another parallel story. Perhaps I was rude.

Perhaps there is another history between people we have loyalties to. Perhaps there is another drama going on that I know nothing about. There are many contexts in which someone who says thank you is not welcome. The silence where "you're welcome" might have been said is not blank. On the contrary, it is filled with communication.

When do we greet? When do we not greet? What are the codes? While it may initially seem that to greet is friendly and to not-greet is rude, there may be multitudes of both. If our distaste for one another is in a context that will not allow the distance we might prefer, the way to greet one another is likely to be vile and dangerous in its saccharine falseness. And if we are very close and I know you are in the kind of pain that needs a witness but not a probing, there are ways to not-greet that offer respect and understanding.

When is language florid and opaque with decoration, and when is it gray concrete?

If you learn the right words to sound empathetic and you follow the script, but the learning has not come up through your bones, it is only a matter of time before a new set of words will be necessary. The right words are no replacement for deeper learning. The worlds will fail, be replaced, be gilded, or be vilified. Their fluidity is necessary while perceptions form. Our fluidity is also necessary while perceptions form.

The way of perceiving is inseparable from the way of describing. The vibe, the atmosphere, and the flavor of the communication create a logic of how to perceive that runs through the context. Architects and interior designers work hard to create spaces with featured metacommunication that tells visitors about who they can be in the building. New perception brings new words. Beyond them both, there is always more going on. The undergrowth of shared cultural signals is always there, monitoring humility, courtesy, oppression, and invitation.

Are you called to speak into this ecology of conversation? And what is brewing that wants to come from your lips? What is the tone? What is the texture of this thing, you will say? If you take a moment and consider the words in their tone and gesture landing in this conversation, what will they bring? What do they do to the alchemy of possibility? If my words land in your ecology of communication, what do they invite? What openings does my communication bring for the conversation? What have I limited? What is closing? What have I nourished, and what have I un-communicated?

It may feel cathartic to "speak your mind" or "say what you want," but both of those are inevitably informed by other ecologies of connection. That thing said or unsaid is not just that and nothing more. Rather, the context into which it happens ripples with the arrival of communication. There is no way to communicate without contextual response and response to response to response. The effects are more than 1st-order; they reach and reach into generations to come at nth-order. Yet, there is little attention to practicing the art of perceiving context, and much gusto is placed on speaking out recklessly.

Taking a stand in a complex interdependent world requires another approach, one with what I have come to call symmathesy or transcontextual mutual learning. This is not a method but a life-long honing and stretching toward sensitizing to perceive differently. Speaking out is important, and doing it with care to context is vital.

In metacommunication, the "meta" is not just communication about communication; it is the implied assumptions of what is possible to communicate in the context and the sonar pings of each expressed and non-expressed interaction against the walls of the expected communication. The algebra teacher may be speaking about formulas—and in doing so, she is also saying something about the math room, the blackboard, the desks and chairs, the history of each student's family, hoping for their child's success. All of these are transmitted in the math teacher's tone of voice and gestures. These implied assumptions are invisibly reiterated into other contexts of the students' lives, like other courses at school, but also family dinners, movies, tech, and window shopping. The implications of the metacommunication in the math class are not measured in how much algebra was learned but are more insidiously submerging into what it is to be in a culture In which algebra is important. The student that tries to succeed is not really succeeding in math but in cultural fitting-in.

Meta is more than self-referential; it is also an evocation of the deeper premises of the context. What is the student learning about learning to be their world? That it is competitive? That it is about pleasing the teacher? That you can fail? It depends on the teacher, the school, the student, the family, the local culture. The meta is there in the way learnings are transformed into a response.

The meta sits in the approach, the attitude, and the complicit understanding of the contexts. It largely is missed in the rush to fix the crises. The communication around solving problems is generally allergic to ecological communication. That is, in the anxiety and urgency to get control and deal with the issues, the familiar strategy perpetuates a habit of flattening, de-vitalizing, and organizing, sorting (even measuring and quantifying)—that which is not to be grasped. The ecology is lost in decontextualized predetermined targets.

I am remembering the Ancient Mariner, who, in a state of total despair, when his crew, the water, the food, and all else was lost—looked over the bow of the ship and saw glowing sea worms in the sea. It is written that he "blessed them unawares." The meta is, as I see it, not the finding of solutions to our multi-systemic emergencies but the difficulty of having to muster gumption, the possibility of experiencing despair and learning to respond in another language of being, another grammar of being alive.

Complete gibberish is starting to make more sense than a good deal of rational debate. In this moment of crumbling social constructs, there is a dire need to see what we have not seen before, to do what we have not done before. There is the need to say what we have not said before. Make up new words . . . lots of them.

The ecology of communication is there to be tended to in every single instant of the day. There is no getting it right; there is only practice and affection for life.

It is written everywhere. In every shape of every leaf. In the bend of every insect's limbs, in the colors and textures of each tree bark, and between your fingers.

It is spoken through every squeak, roar, and word salad. Every crack of thunder and every song of every river and creek is saying it.

It is in the rhythm of the waves crashing, the seasons passing, and the birth of the fawns in spring. The heart of each animal keeps the time. The crickets and cicadas hold the night's hum. The wind marks the desert in stripes and ridges. The footsteps of a horse, a family member. A returning hunger.

The gestures are wide horizons or jungles teeming with greens. They are flung high into murmuration of birds and down into the filigree of the undergrowth fungi. You shift your eye and blush . . . the cat twitches its tail while a city skyline reaches upward and clutches right angles in eager corners.

Tone is key. The swish of breeze high in the birch trees contrasts with the sound of an earthquake growling from the deep. The cheerful din of a meal with friends is paired with the private hell of angry silence between couples. The flattening inflection of authorized media voices tells listeners that credibility is the soundscape of dry paste, wiping clean any slime of uncertainty.

There is, within all of this, another realm that is beyond communication. That realm is sacred and unexplained. To force an explanation would be to violate a vital communion. There is the way a piece of music can move us to tears and the way a sudden glimpse of insight arrives.

There is love.

Fig. 53. Leslie Thulin. (2022). *Mending.* [digital art].

INTEGRITY

Integrity is the art of navigating those moments in which no pre-scripted rules apply. To meet complexity, to make new sense of it, requires willingness to go forward without someone else's instructions. It means not justifying continued destruction or limiting what is possible by pointing to history, or worse, "human nature," whatever that is.

The rigor of this is daunting. Comparing and contrasting inner interpretations, experiences—listening and watching with wide-angle perception. I see it as a muscle to train in receiving information about the vast interrelational consequences of each action and non-action. (And, the action not taken is also an action.) It means staying alert, paying attention, and resisting the itch to rest in the familiar.

Integrity has to do with knowing that I will show up with all I can muster.

Combining

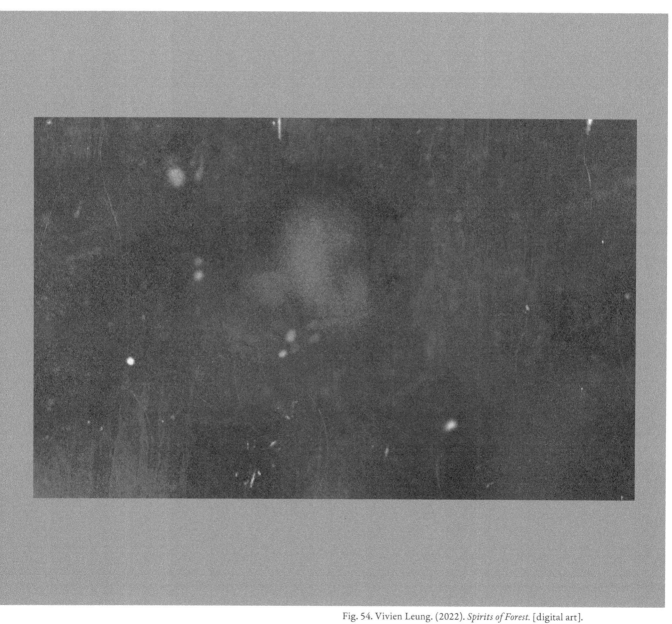

Fig. 54. Vivien Leung. (2022). *Spirits of Forest*. [digital art].

Combining

SOMETHING HAS TO MATTER

What matters continues.

Surface distractions,
And day-to-day transactions,
are loud
the front part of our lives,
but what matters stays below.

Something is under us.
Like a drum beating through us.

Mattering mattering mattering.

Perhaps we have not even noticed what matters.

I thought I heard him say,
No one has time to wait for the moon.

That which matters takes time to tend.
Time to remember, time to become so intrinsic it is in our bones,

It takes air, and water, and earth, and memory.
It takes hard work,
Day in
Day out,
Generation after generation.

That which matters is unsayable in words.
And is communicated through us.

It is the grain in our wood. The hum in the song.
The patterns of our lives bounce around in a ricochet room.
Repeating and responding in the same humming grain.

Infinity matters in a world of faster, cheaper,
more, more, more ...

                              The moon has time.
                             And this moment,
                                   this meal,
                                   this walk,
                                   this work,
                              this cup of tea
                        is never to be repeated.

                                It is precious.
                       It matters what continues.
             It is the code through which all else will be.

                                    It matters.

COMBINING

Life is made in relationshipping.

Communicating in shared nutrients, as trees do, is a beautiful way to think of how our communion might also be.

Life is an enormous conversation.

HOME

If place makes me
Then maybe I don't exist.
And if the memory of redwood dust is enough to
find outlines
Then maybe I am still nine.
Home.
Is a long story,
And I am the ink, the song, the characters.
I am the shoes, the dusty windowsill, and the
rain outside.
I am the moment where you wonder if the threads
will come together.

Combining

Let us un-thing it, un-name it, un-it "it," finding warmth and relief in possibilities that naming and thing-ing and it-ing obscure.

This collection of words and ideas is a courtship with the tenderness of life as it is life-ing. The scaffolding of control is seamlessly wrought into the logic of day-to-day living—a tenderness such as this is illegible. All the language that tries to capture it has the filth of the capturing on it. There is nothing to take, have, gain, or lose. This book is not about systems change. It is about looking around one day and realizing that ABSOLUTELY EVERYTHING has been altered into an entirely different approach.

The communication must be weird . . . unexpected . . . poetic . . . and irritating and fucked up and staggeringly beautiful. Some things cannot be said in the tones and vocabularies that hold the shape of the familiar ideas. There is a need to get freaky—expand the brackets of what is possible to explore.

This is not in you or me but in what is implicit between us. If the implicit is a warehouse of scripts built on old scripts—it will be highly likely that the same-old-same-old will continue.

Speaking a new and unrehearsed sentence brings a strange texture to your mouth. Moving with a new and unrehearsed rhythm brings an unfamiliar shape and gesture to your body. This matters. Making new communication possible is wild. Domesticated ideas stand aside in nice rows, tidy boxes, polished language, and preferred grammars.

No, this is not safe, but then again, safe is not safe either.

CUPPED HANDS

There is a dark, smooshy substance with small bits; some have woven patterns in them, others are spikey, some are globular. They are in exciting colors and move in differing tempos. They appear in what could be described as a song. I am up to my knees in this goop.

Strange that it has been there all my life, and I just noticed it. All this time, I assumed I was walking in shoes, on roads and paths, on solid ground, held by gravity.

It tickles, so I reach down with cupped hands and bring a scoopful to my face. This is wild and necessary.

This book does not want to be a book; it wants to be a meadow.

Life wants it to live in you and me, continuing to shift us wordlessly. So, I am inviting a place for the notions to be on the page and, at the same time, to continue re-impressing and expression-ing in the underground.

The ideas—and the words around them—are alive. I do not want to peel them off the pavement of practicality to lay them flat in definitions, methodologies, curricula, tools, and other such monstrosities. Like underwater bioluminescent organisms, they are flittering, wiggling alight in contact as they gather and touch other life. There is nothing sterile to say; it is life that is speaking.

Now I see—the time is at hand to carefully tend the radical noticing of how so many contexts of life have been intersteeping for the last several thousand years to form a fusion that attaches survival to extinction.

Readying ourselves and each other to be in a world that is stretching, ripping, opening, closing, releasing, and gushing into new forms is another thing altogether.

Fig. 55. (pp. 370-382). Mats Qvarfordt & Trevor Brubeck. (2020-2023). *Test Print Collection*. [Handprinted Wallpaper.].
@Handtryckta Tapeter Långholmen

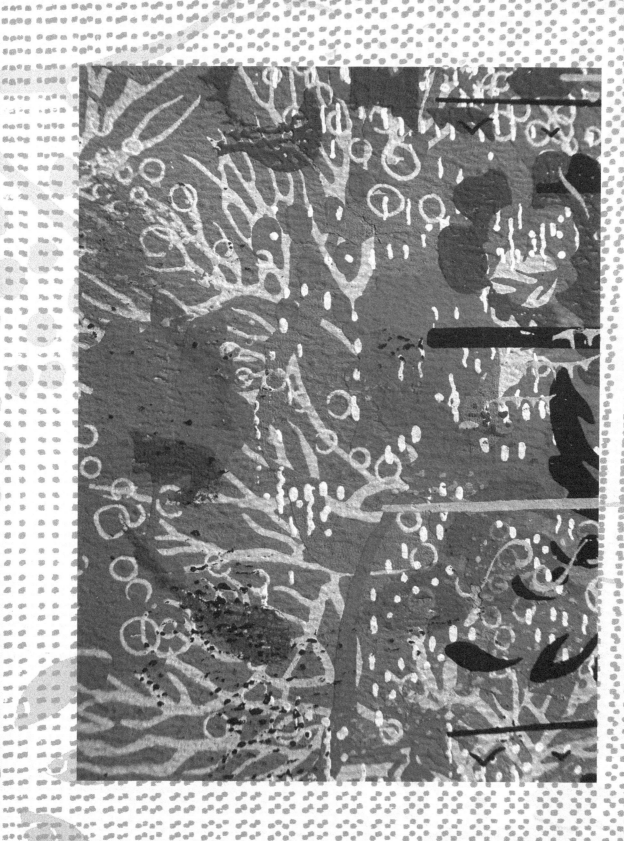

References and Contexts

Amend, A. (2019, September 29). First as tragedy, then as fascism: Alex Amend. *The Baffler*. https://thebaffler.com/latest/first-as-tragedy-then-as-fascism-amend

Arentsen, T., Raith, H., Qian, Y., Forssberg, H., & Diaz Heijtz, R. (2015). Host microbiota modulates development of social preference in mice. *Microbial Ecology in Health and Disease*, 26, 29719. https://doi.org/10.3402/mehd.v26.29719

Bateson, B. (2009). *William Bateson, Naturalist: His Essays and Addresses Together with a Short Account of His Life*. Cambridge University Press.

Bateson, G. (1979). *Nuclear Armament as Epistemological Error* – Letters to the California Board of Regents. Zero, 3, 34–41.

Bateson, G. (1980). *Simple Thinking* [Audio Recording] [Lecture]. Esalen Institute.

Bateson, G. (1991). *A Sacred Unity: Further Steps to an Ecology of Mind* (R. E. Donaldson, Ed.). Cornelia & Michael Bessie Book.

Bateson, G. (2000). *Steps to an Ecology of Mind: Collected Essays in Anthropology, Psychiatry, Evolution, and Epistemology* (1st ed.). University of Chicago Press.

Bateson, G. (2002). *Mind and Nature: A Necessary Unity*. Hampton Press.

Bateson, G. & Bateson, M. C. (1988). *Angels Fear: Towards an Epistemology of the Sacred*. Bantam.

Bateson, M. C. (1972). *Our Own Metaphor: A Personal Account of a Conference on the Effects of Conscious Purpose on Human Adaptation* (1st ed.). Alfred A. Knopf.

Bateson, M. C. (1995). *Peripheral Visions: Learning Along the Way*. HarperCollins.

Bateson, N. (2016). *Small Arcs of Larger Circles: Framing Through Other Patterns*. Triarchy Press.

Bateson, N. (2017, May 28). *Warm Data*. Hacker Noon. https://hackernoon.com/warm-data-9f0fcd2a828c

Bateson, N. (2017). Liminal leadership. *Kosmos: Journal for Global Transformation*. https://www.kosmosjournal.org/article/liminal-leadership/

Bateson, N. (2017). The Era of Emergency Relocation-A Transcontextual Perspective. *Fokus på familien*, 45(2), 82-98. https://doi.org/10.18261/issn.0807-7487-2017-02-02

Bateson, N. (2018). # MeToo is complex. *Explorations in Media Ecology*, 17(1), 71-76. https://doi.org/10.1386/eme.17.1.71_7

Bateson, N. (2019). I Want You to Want Me to Want You [and vice versa]: The Simple Complexity of Sexual Consent. *Journal of Design and Science*. https://jods.mitpress.mit.edu/pub/hqglqga6

Bateson, N. (2020). Preface. In: Salvatore D'Amore ed., *The Challenges of Today's Families: A Systemic Approach to Family Relationships* (pp. 13-15). De Boeck Superior. https://doi.org/10.3917/dbu.damor.2020.01.0013

Bateson, N. (2021). Aphanipoiesis. *Journal of the International Society for the Systems Sciences,* 65(1), 1–16. https://journals.isss.org/index.php/jisss/article/view/3887

Bateson, N. (2022, Summer). Reunion. New words to hold the invisible world of possibility: warm data, symmathesy and aphanipoiesis. *Unpsychology Issue 8: An Anthology of Warm Data,* (8), 12–15. https://unpsychology.substack.com/p/new-words-to-hold-the-invisible-world

Bateson, N. (2022, September 21). An essay on ready-ing: Tending the prelude to change. *Systems Research and Behavioral Science,* 39(5), 990–1004. https://onlinelibrary.wiley.com/doi/abs/10.1002/sres.2896

Bateson, W. (1888, September 2). *William Bateson to Anna Bateson* (MS Add.8634/A.29:4.1r-A.29:4.4v). Cambridge University Library, Department of Manuscripts and University Archives. https://cudl.lib.cam.ac.uk/view/MS-ADD-08634-A/1557

Browning, R., Markham, E. (1856). *Men and Women.* Ticknor and Fields.

Clayton, A. (2020). How eugenics shaped statistics. *Nautilus.* https://nautil.us/issue/89/the-dark-side/how-eugenics-shaped-statistics

Clayton, P., Archie, K. M., Sachs, J., Steiner, E., Robinson, K. S., & Ramphele, M. (2021). Ubuntu: The Dream of a Planetary Community. In *The New Possible: Visions of Our World Beyond Crisis.* [Essay], Cascade Books.

Erdrich, L. (1998). *The Antelope Wife: A Novel.* HarperFlamingo.

Foerster, H. v. (2003). *Understanding Understanding: Essays on Cybernetics and Cognition.* Springer. https://doi.org/10.1007/0-387-21722-3_12

Hardin, G. (1974). Commentary: Living on a lifeboat. *BioScience,* 24(10), 561–568. https://doi.org/10.2307/1296629

Jameson, F. (1994). *The Seeds of Time.* Columbia University Press.

Kauffman, S. (1995). *At Home in the Universe: The Search for Laws of Self-Organization and Complexity.* Oxford University Press.

Korzybski, A. (1994). *Science And Sanity: An Introduction to Non-Aristotelian Systems and General Semantics* (Fifth). Institute of General Semantics International Non-Aristotelian Library Publishing Company, 1933.

Kubrick, S. (Director). (1964). *Dr. Strangelove or: How I Learned to Stop Worrying and Love the Bomb* [Film]. Hawk Films.

Le Guin, U. K. (1984). *Left Hand of Darkness.* ACE Science Fiction Books.

Locher, F. (2013). Les Pâturages de la Guerre Froide : Garrett Hardin et la « tragédie des communs ». *Revue d'Histoire Moderne et Contemporaine,* 60–1(1), 7–36. https://doi.org/10.3917/rhmc.601.0007

Margulis, L., & Sagan, D. (1995). *What is Life?* Simon & Schuster.

Mascolo, M. F. (2016, December 4). The False Lure of Objectivity in Psychology. *Psychology Today.* Retrieved January 7, 2023, from https://www.psychologytoday.com/us/blog/values-matter/201612/the-false-lure-objectivity-in-psychology

Maturana, H. R., & Varela, F. J. (1980). *Autopoiesis and Cognition: The Realization of the Living.* Springer Netherlands.

McLuhan, M. (1964). *Understanding Media: The Extensions of Man.* Routledge & Kegan Paul.

Mead, Margaret. (1928). *Coming of Age in Samoa.* W. Morrow & Company.

Morim, A. G., Demarchi, A., Lima, M. R. P., & Omim, S. (2013). Uma conversa sobre a ecologia da mente: Entrevista com Nora Bateson. *Enfoques,* 12(1), 266-283.

Peirce, C. S., Houser, N., & Kloesel, C. (1998). *The Essential Peirce. Selected Philosophical Writings: 1893-1913* (Peirce Edition Project Ed., Vol. 2). Indiana University Press.

Rosen, J. (2012). Preface to the Second Edition: The Nature of Life. In R. Rosen, *Anticipatory Systems: Philosophical, Mathematical, and Methodological Foundations* (2nd ed., pp. xi-xiv). Springer. Retrieved from http://doi.org/10.1007/978-1-4614-1269-4

Russell, B. (1908, July). Mathematical Logic as Based on the Theory of Types. *American Journal of Mathematics,* 30(3), 222-262. JSTOR. https://doi.org/10.2307/2369948

Somé, M. P. (1999). *The Healing Wisdom of Africa: Finding Life Purpose Through Nature, Ritual, and Community.* Penguin Publishing Group.

Von Foerster, H. (1984). *Observing Systems.* Intersystems Publications.

Watts, A. (1995). *The Tao of Philosophy: The Edited Transcripts.* C.E. Tuttle.

Whitehead, A., & Russell, B. (1997). *Principia Mathematica to 56* (2nd ed., Cambridge Mathematical Library). Cambridge University Press. doi:10.1017/CBO9780511623585

Whitehead, A. N. (1925). *An Enquiry Concerning the Principles of Natural Knowledge.* University Press.

Art Credits

Fig. 1. Nora Bateson (concept & text), Rachel Hentsch & Vivien Leung (design), Leslie Thulin (research). (2023). Moths & Butterflies Spread. [digital art]..6
Fig. 1a. imageBROKER. (n.d.). Eyed Hawk-Moth, Smerinthus ocellata. [photograph]. Retrieved from stock.adobe.com. 6
Fig. 1b. dinar. (n.d.) From The biggest butterfly in the world Attacus atlas close-up. [photograph]. Retrieved from stock.adobe.com. 6
Fig. 1c. mramsdell1967. (2019). Butterfly 2019-119 / Two Blue Morpho Butterfly. [photograph]. Retrieved from stock.adobe.com. 6
Fig. 1d. Stockgalp. (n.d.) Butterflies in the subtropical region of MASHPI rainforest in Ecuador. [photograph]. Retrieved from stock.adobe.com. 7
Fig 2. Vivien Leung. (2023). #EEAAO Madness. [digital art]. 18
Fig. 3. Vivien Leung. (2023). Where We Listen. [digital art]..............19
Fig. 4. Nora Bateson. (2023). More Than Blobs. [oil on canvas]......... 38
Fig. 5. Vivien Leung. (2023). Blooming. [digital art]. 52
Fig. 6. Vivien Leung. (2023). Ocean Breeze. [digital art]................ 53
Fig. 7. Maria Sibylla Merian. (1701-1705). "A pineapple surrounded by cockroaches." [Watercolor and bodycolor on vellum]. Image from the British Museum. London, United Kingdom. 56
Fig. 8. Dorothea Graff or Johanna Herolt after Maria Sibylla Merian. (1701-1705). "Tarantulas, one attacking a hummingbird, spiders and ants, on a guava tree." [Image of Watercolor and bodycolor on vellum]. Image from the British Museum. London, United Kingdom.......................... 57
Fig. 9. Rachel Hentsch. (2023). Assholery. [digital art]................. 62
Fig. 10. Vivien Leung. (2023). Combining. [digital art]. 63
Fig. 11. Vivien Leung. (2023). Bellflower. [digital art].................. 64
Fig. 12. Rachel Hentsch. (2023). Connective Tissue. [digital art]........ 67
Fig. 13. Rachel Hentsch. (2022). Expressing. [digital art].................68
Fig. 14. (pp. 78-85). Rachel Hentsch (design, composition), with Nora Bateson (photography, poetry text). (2023). Stretching Edges. [digital art].78
Fig. 15. Vivien Leung. (2023). Fainted Light. [digital art]................88
Fig. 16. Vivien Leung. (2023). What We Perceive. [digital art]...........89
Fig. 17. Rachel Hentsch. (2023). Fantastic Six Blobs. [digital art]. 92
Fig. 18. Leslie Thulin. (2022). Always Moving. [digital art]............. 98
Fig. 19. Leslie Thulin. (2023). Together. [digital art]. 102
Fig. 20. (pp. 110-119). Rachel Hentsch. (Collection: curation, design, layout, composition). (2023). Crossing Borders: Families in Motion. [digital art].112
Fig. 20a. Mats Qvarfordt & Trevor Brubeck. (2020–2023). Untitled. [Handprinted wallpaper test sheets]. 112
Fig. 20b. Nora Bateson & Vivien Leung. (2022). Belly. [Gansai paint and ink on Sumi-e paper].. 113
Fig. 20c. Nora Bateson. (2023). Blobs Collection: Golden. [acrylic on canvas]. 114
Fig. 20d. Nora Bateson. (2022). Blob Pathway. [ink on paper].......... 115

Fig. 20e. Nora Bateson. (2023). Family. [oil on canvas]................ 116
Fig. 20f. Mats Qvarfordt & Trevor Brubeck. (2020–2023). Untitled. [Handprinted wallpaper test sheets]............................... 117
Fig. 20g. Nora Bateson. (2023). Biology Blob. [oil on canvas].......... 117
Fig. 20h Mats Qvarfordt & Trevor Brubeck. (2020–2023). Untitled. [Handprinted wallpaper test sheets]............................... 117
Fig. 20i. Nora Bateson. (2023). Blob in Blob in Blob. [oil on canvas]..... 118
Fig. 20j. Mats Qvarfordt & Trevor Brubeck. (2020–2023). Untitled. [Handprinted wallpaper test sheets]............................... 118
Fig. 20k. Nora Bateson. (2023). Community Blobs. [oil on canvas]...... 119
Fig. 20l. Nora Bateson. (2022). Blob History. [ink on paper]............ 120
Fig. 20m. Mats Qvarfordt & Trevor Brubeck. (2020–2023). Untitled. [Handprinted wallpaper test sheets]............................... 121
Fig. 21. Rachel Hentsch. (2023). Veil. [digital art]..................... 128
Fig. 22. Vivien Leung. (2022). Play. [digital art]...................... 138
Fig. 23. Vivien Leung. (2023). Moving. [digital art]................... 139
Fig. 24. Vivien Leung. (2023). Materials Forming. [digital art].......... 140
Fig. 25. Rachel Hentsch. (2021). Aphanipoiesis One. [digital art]....... 148
Fig. 26. Rachel Hentsch. (2021). Aphanipoiesis Two. [digital art]....... 151
Fig. 27. Vivien Leung. (2023). Mama Hands. [digital art]............. 157
Fig. 28. Vivien Leung. (2022). Untitled. [digital art]. 164
Fig. 29. Vivien Leung. (2023). Searching. [digital art]................. 174
Fig. 30. Vivien Leung. (2023). Life Spiraling Art. [digital art].......... 175
Fig. 31. Vivien Leung. (2023). Mama Blobs. [digital art].............. 176
Fig. 32. (pp. 186-201). Vivien Leung (design and handwriting), Nora Bateson (poetry text & concept), Mats Qvarfordt & Trevor Brubeck @Handtryckta Tapeter Långholmen (wallpaper background), Rachel Hentsch (graphics), Leslie Thulin (graphics). (2023). The Meadow Verse. [digital art on images of handprinted wallpaper]... 186
Fig. 33. Vivien Leung. (2022). Untitled. [digital art]. 210
Fig 34. Vivien Leung. (2023). Bellflower. [digital art]. 211
Fig. 35. Rachel Hentsch. (2023). Combining Elements. [photocollage & digital art]... 230
Fig. 36. Rachel Hentsch. (2023). Water Squares. [digital art].......... 231
Fig. 37. (pp. 248-255). Vivien Leung (drawing & design), Nora Bateson (concept and text), Leslie Thulin (text edit), Phillip Guddemi (text edit). (2023). Glossary. [digital art]....................................... 248
Fig. 38. Rachel Hentsch. (2023) To Live in Another Way. [digital art]. .. 260
Fig. 39. Rachel Hentsch. (2023). Meet Not Match. [digital art]. 281
Fig. 40. Rachel Henstch. (2021). Not Just That And Nothing More, version 1. [digital art].. 294
Fig. 41. Rachel Henstch. (2021). Not Just That And Nothing More, version 2. [digital art].. 295
Fig. 42. Gregory Bateson. (1966). Porpoises. [photograph]............ 300

Fig. 43. Vivien Leung. (2023). Belly Spinning. [digital art].............. 305
Fig. 44. Alexei Jawlensky. (1917). Mystical Head: Galka. [Image of oil and pencil on tan textured cardboard]. Image from the Norton Simon Museum, The Blue Four Galka Scheyer Collection 306
Fig. 45. Ursus Wehrli. (2004). Kunst aufräumen. [Image of painting]. Copyright © Kein und Aber Verlag Königstein i. Ts. 307
Fig. 46. (pp. 308-315). Rachel Hentsch (design & concept), with Nora Bateson (Concept & Text) and Leslie Thulin (Concept). (2023). Transcontextual. [digital art]. 308
Fig. 47. Leslie Thulin. (2023). Weaving. [digital art]. 318
Fig. 48. Leslie Thulin. (2023). Scraping. [digital art]. 318
Fig. 49. Leslie Thulin. (2023). Cracking Frequency. [digital art]......... 318
Fig. 50. Leslie Thulin. (2023). Dawn. [digital art]..................... 328
Fig. 51. Vivien Leung. (2023). Caring. [digital art]..................... 346
Fig. 52. Vivien Leung. (2023). Transmitting Vibe. [digital art].......... 347
Fig. 53. Leslie Thulin. (2022). Mending. [digital art].................. 360
Fig. 54. Vivien Leung. (2022). Spirits of Forest. [digital art]. 363
Fig. 55. (pp. 370-382). Mats Qvarfordt & Trevor Brubeck. (2020-2023). Test Print Collection. [Handprinted Wallpaper.]. @Handtryckta Tapeter Långholmen 370

Contents (In Alphabetical Order)

About the Author 393
About the Book 393
Acknowledgments xiii
Affection for Life 141
A Letter to My Imagination 228
An Ecology of Assholes 220
Aphanipoiesis 145
A Pineapple Surrounded by Cockroaches 56
Art Credits 386
Bacteria 263
Building an Arc 265
Common Sense Is Sense-making in the Commons 273
Communing 54
Contents ix
Cracks and Fissures 123
Creature 202
Crossing Borders: Families in Motion 113
Cupped Hands 369
Decontextualized 306
Dedication vii
Divided We Fall Together 216
Do Something 9
Ecology of Communication 353
Eggs Are Time 12
Every Hole Is a Story 90
Family Is Where We Live 316
Finding a Way 69
For You 217
Freak Out and Freak In 267
Frequency 278
Frost 125
Glossary 248
Hallway of Hallways 39
Harvest 350
Home 367

How Do You Pack?...................218
I Am a Crayon.....................206
Ideas Are Their Stories 256
I Fear a Fear of Fear.............. 122
I Love You........................91
Integrity......................... 361
In the Fire.......................282
Introduction........................1
It's a Gap 177
It's Fantastic.................... 99
I Want You to Want Me to Want You ... 331
Juicy.............................44
Just Sing.........................124
Kinky 185
Life Is Art 109
Liminal Leadership................ 233
Listening to the Listeners 183
Lurking Monster 271
Mama Now 42
Marrow 209
Meeting Double Binds in the Polycrisis............... 319
Meet Not Match 21
Minutiae of the Day 280
Moths and Butterflies..............7
Moving Edges......................40
New Blank Document 212
Nocturnal.........................264
Noticing..........................184
Now 94
One Thing.........................65
Possibility 13
Predatory Skills 349
References and Contexts 383
Rejection.........................223
Reunion...........................111

Sacred Communication................................219
Salt and Iron...259
Seasons Change Everything While Breaking Nothing.....109
Self Portrait..86
Silences...348
Simultaneously Implicating 103
Slow Truth..329
Somehow..77
Something Has to Matter 364
Something New..225
Stretching Edges..78
Surreal...304
Swerving 226
Symmathesy...110
Tacit...126
Tarantulas, Hummingbird, Spiders and Ants............ 57
Tearing and Mending...................................283
The Caramels of Autumn 45
The Cringe...222
The Meadow-Verse 186
Theory is Beautiful 257
The Reasons 258
(There Is No Script) 279
The Rubric...269
The Zombie Caterpillar................................261
Time in Winter Is Underground........................207
To Live in Another Way..................................5
To Live It...352
Tone..75
Traveling on a Paved Road 76
Two Bad Questions.....................................224
Unbreakable..301
Uncut...55
Un-pick-apart-able 47
Unsilent...208

Combining

Untamed 144
Urgent Mud 142
What am I Not Able to Receive? 303
What I Learned 96
What Is Sanity? 272
What Is Submerging? 129
Where Is The Edge of Me? 59
Where Prose Stumbles 3
Who-New? 302
Wild 127
Without Going Blank 93
Without Shields (The Voice of Change Is Changing) 66
Words to be Careful With 241
Yes 215

About the Author

Nora Bateson is an award-winning filmmaker, artist, international lecturer, research designer, author, as well as Founder and President of the International Bateson Institute. She is the founder and creator of the concept "Warm Data" and the practices of the Warm Data Lab and People Need People Online.

Nora wrote, directed, and produced the documentary, *An Ecology of Mind, A Daughter's Portrait of Gregory Bateson.* Her work brings the fields of biology, cognition, art, anthropology, psychology, and information technology together into a study of patterns in the ecology of living systems. Her book, *Small Arcs of Larger Circles,* published by Triarchy Press, UK, 2016 is a revolutionary personal approach to the study of systems and complexity. Nora and her husband, Mats Qvarfordt, share a household of handicraft, music, art, and cooking.

About the Book

Like a mole in the garden of our times, velvet, sharp-eyed and half underground, Nora Bateson's *Combining* undermines our most precious certainties, offers nourishment, unseen circulation, surprises, mounds, chambers and drainage tunnels for our thinking, living and relating.

Using multiple artistic and literary forms, resisting always the urgent request for answers, checklists and action points, Nora's work is the embodiment of the Warm Data approach that she teaches. It is the most telling example of what she has named aphanipoiesis: that prolific tendency (characteristic of all living systems) for things to come together or coalesce unseen towards vitality.

If ever a book could work overnight, discreetly, very quietly, to change our minds, it's *Combining*.

Triarchy Press